MW01042882

student study

ART NOTEBOOK

Introduction to BIOLOGY

SYLVIA S. MADER

WCB Wm. C. Brown Publishers
Dubuque, Iowa • Melbourne, Australia • Oxford, England

Book Team

Editor *Carol J. Mills*
Developmental Editor *Diane Beausoleil*
Production Editor *Catherine S. Di Pasquale*
Designer *Anna C. Manhart*
Art Editor *Joseph P. O'Connell*
Photo Editor *Carrie Burger*
Permissions Coordinator *Vicki Krug*

Wm. C. Brown Publishers
A Division of Wm. C. Brown Communications, Inc.

Vice President and General Manager *Beverly Kolz*
Vice President, Publisher *Kevin Kane*
Vice President, Director of Sales and Marketing *Virginia S. Moffat*
National Sales Manager *Douglas J. DiNardo*
Marketing Manager *Craig Johnson*
Advertising Manager *Janelle Keeffer*
Director of Production *Colleen A. Yonda*
Publishing Services Manager *Karen J. Slaght*
Permissions/Records Manager *Connie Allendorf*

Wm. C. Brown Communications, Inc.

President and Chief Executive Officer *G. Franklin Lewis*
Corporate Senior Vice President, President of WCB Manufacturing *Roger Meyer*
Corporate Senior Vice President and Chief Financial Officer *Robert Chesterman*

The credits section for this book begins on page 91 and is considered an extension of the copyright page.

A Times Mirror Company

ISBN 0-697-23170-4

Printed in the United States of America by Wm. C. Brown Communications, Inc., 2460 Kerper Boulevard, Dubuque, IA 52001

10 9 8 7 6 5 4 3 2 1

TO INSTRUCTORS AND STUDENTS

This Student Study Art Notebook is free with a new textbook to all students and can be used to take notes during lectures. On each notebook page, there are two figures (sometimes one, sometimes three) faithfully reproduced from the original textbook figure. Each figure also corresponds to each of the 100 acetates available to instructors with adoption of the text.

The intention is to place a copy of the transparency acetate art in front of students (via the notebook) as the instructor uses the overhead during lectures. The advantage to the student is that he/she will be able to see all labels clearly, and take meaningful notes without having to make hurried sketches of the acetate figure.

The pages of the Art Notebook are perforated and three-hole punched, so they can be removed and placed in a personal binder for specific study and review, or to create space for additional notes.

DIRECTORY OF NOTEBOOK FIGURES

TO ACCOMPANY

INTRODUCTION TO BIOLOGY

BY SYLVIA S. MADER

Ionic Reaction
Figure 2.2

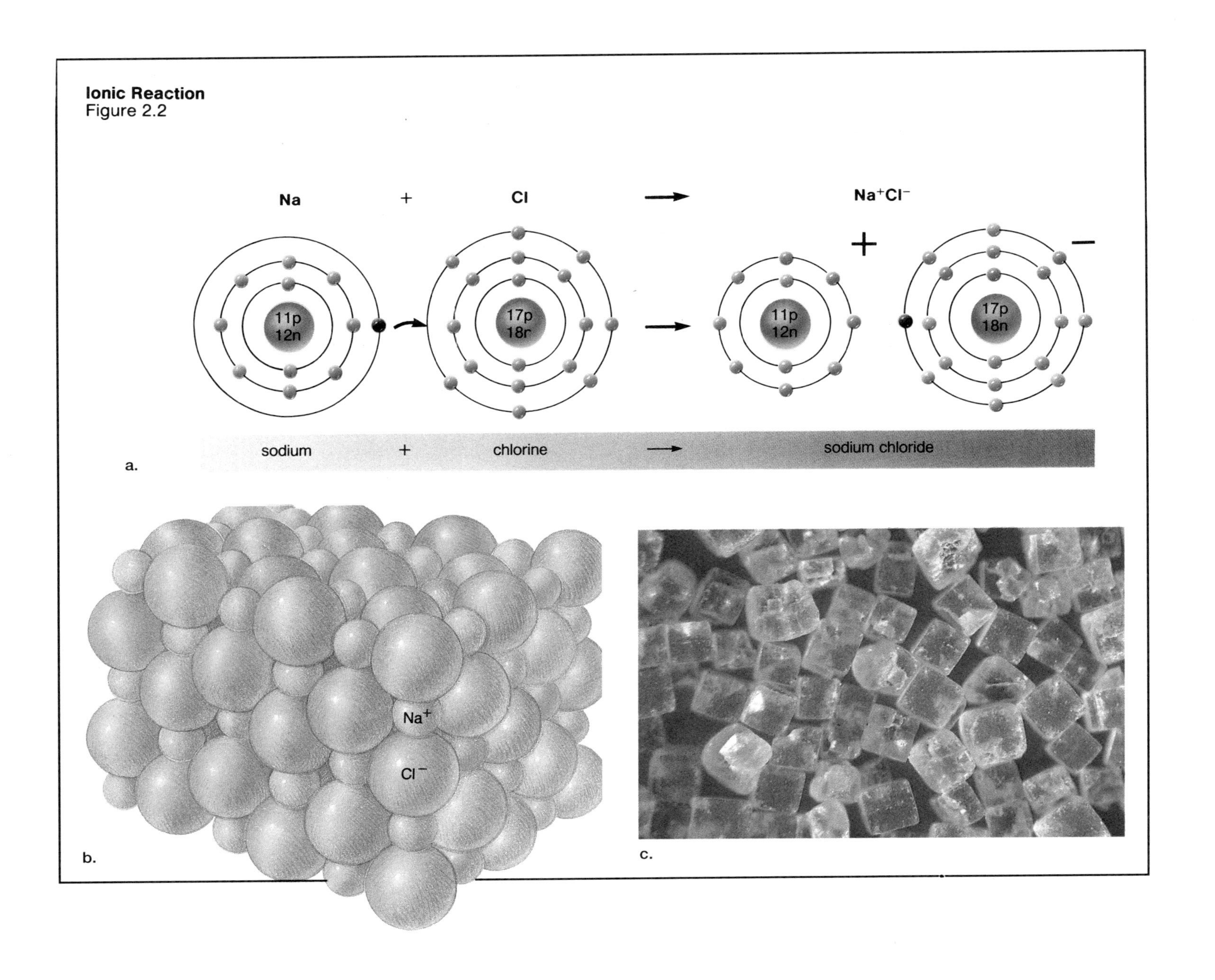
Na
+
Cl
→
Na+Cl−
11p
12n
17p
18r
11p
12n
17p
18n
+
−
a.
sodium
+
chlorine
→
sodium chloride
Na+
Cl−
b.
c.

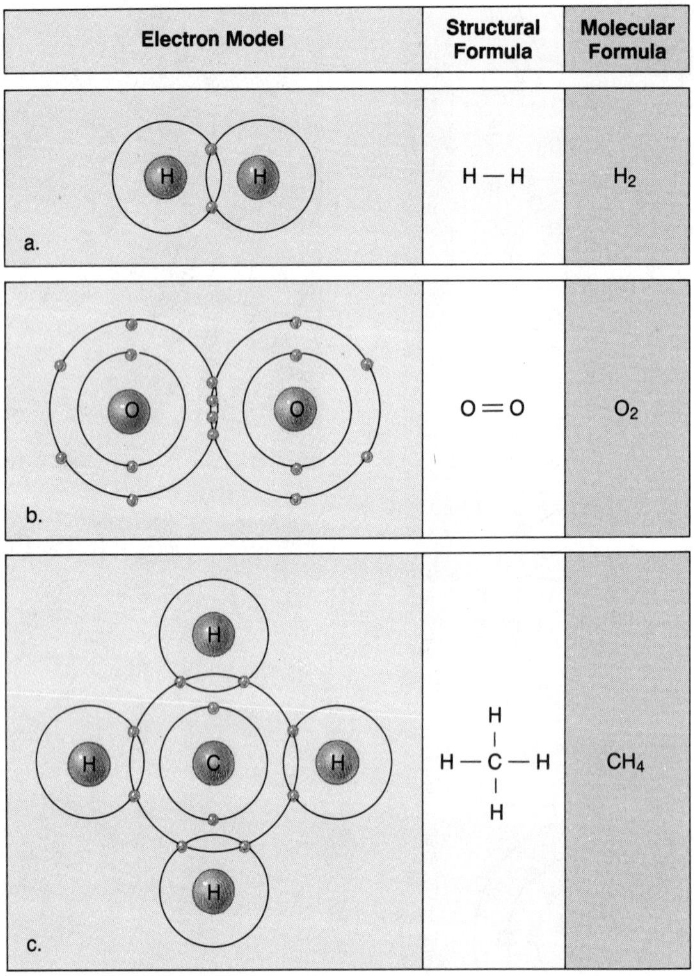

Electron Model	Structural Formula	Molecular Formula
a. (H H)	H — H	H_2
b. (O O)	O ═ O	O_2
c. (H, H, C, H, H)	H — C — H, with H above and below C	CH_4

Covalent Bonding
Figure 2.3

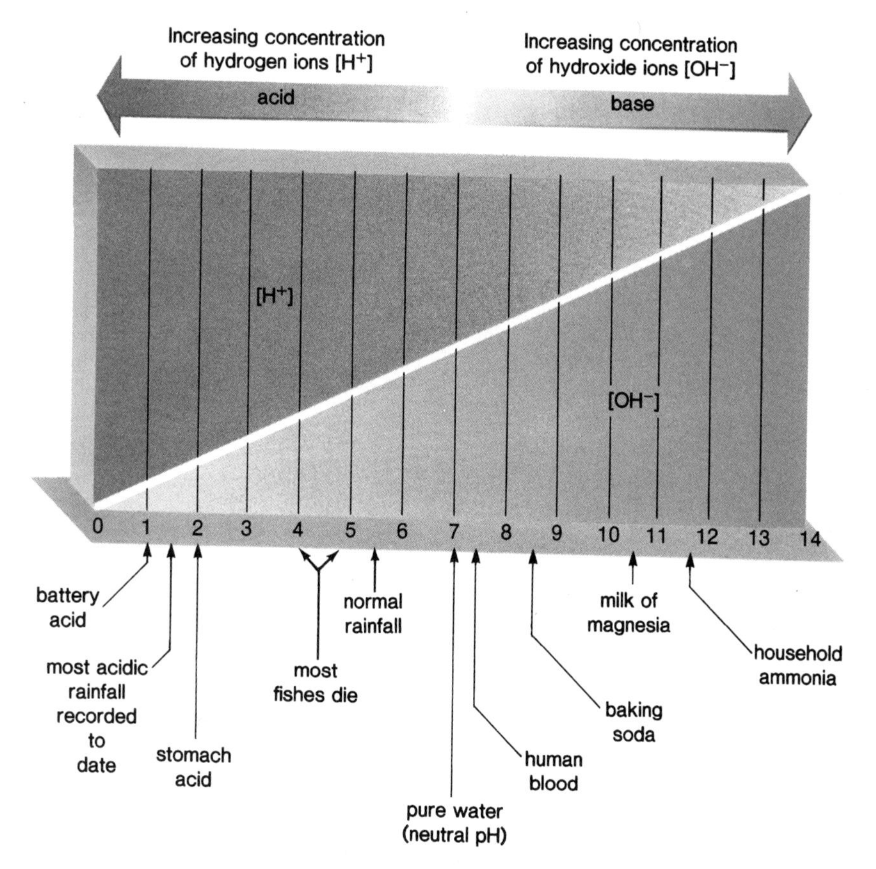
Increasing concentration of hydrogen ions [H+]
Increasing concentration of hydroxide ions [OH−]
acid
base
[H+]
[OH−]
0 1 2 3 4 5 6 7 8 9 10 11 12 13 14
battery acid
most acidic rainfall recorded to date
stomach acid
most fishes die
normal rainfall
pure water (neutral pH)
human blood
baking soda
milk of magnesia
household ammonia

The pH Scale
Figure 2.5

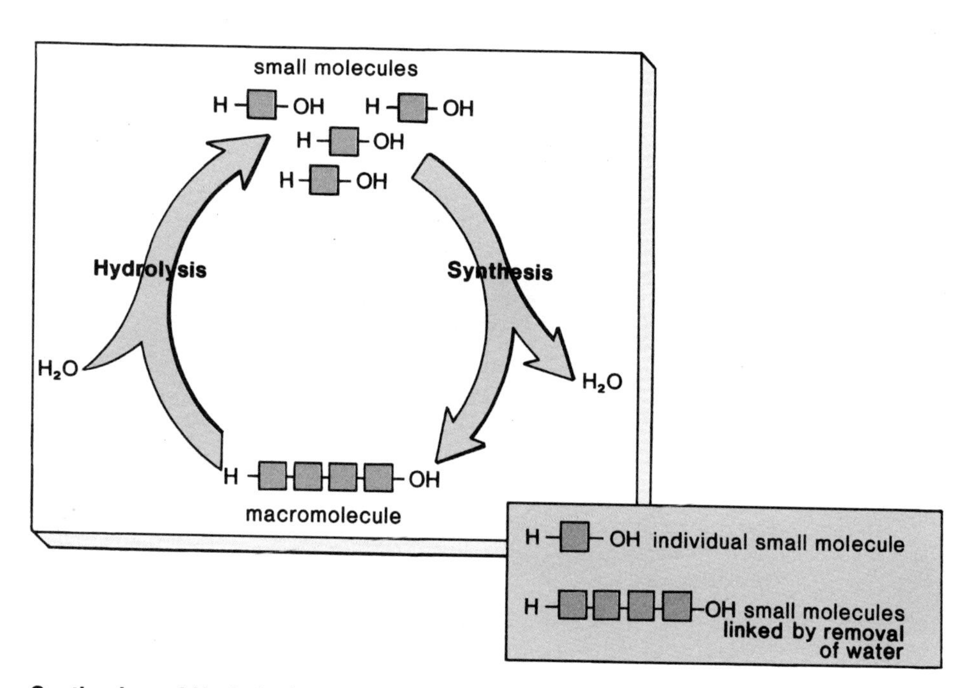
small molecules
H – OH
Hydrolysis
Synthesis
H₂O
H₂O
macromolecule
H – OH individual small molecule
H – OH small molecules linked by removal of water

Synthesis and Hydrolysis of Macromolecules
Figure 2.6

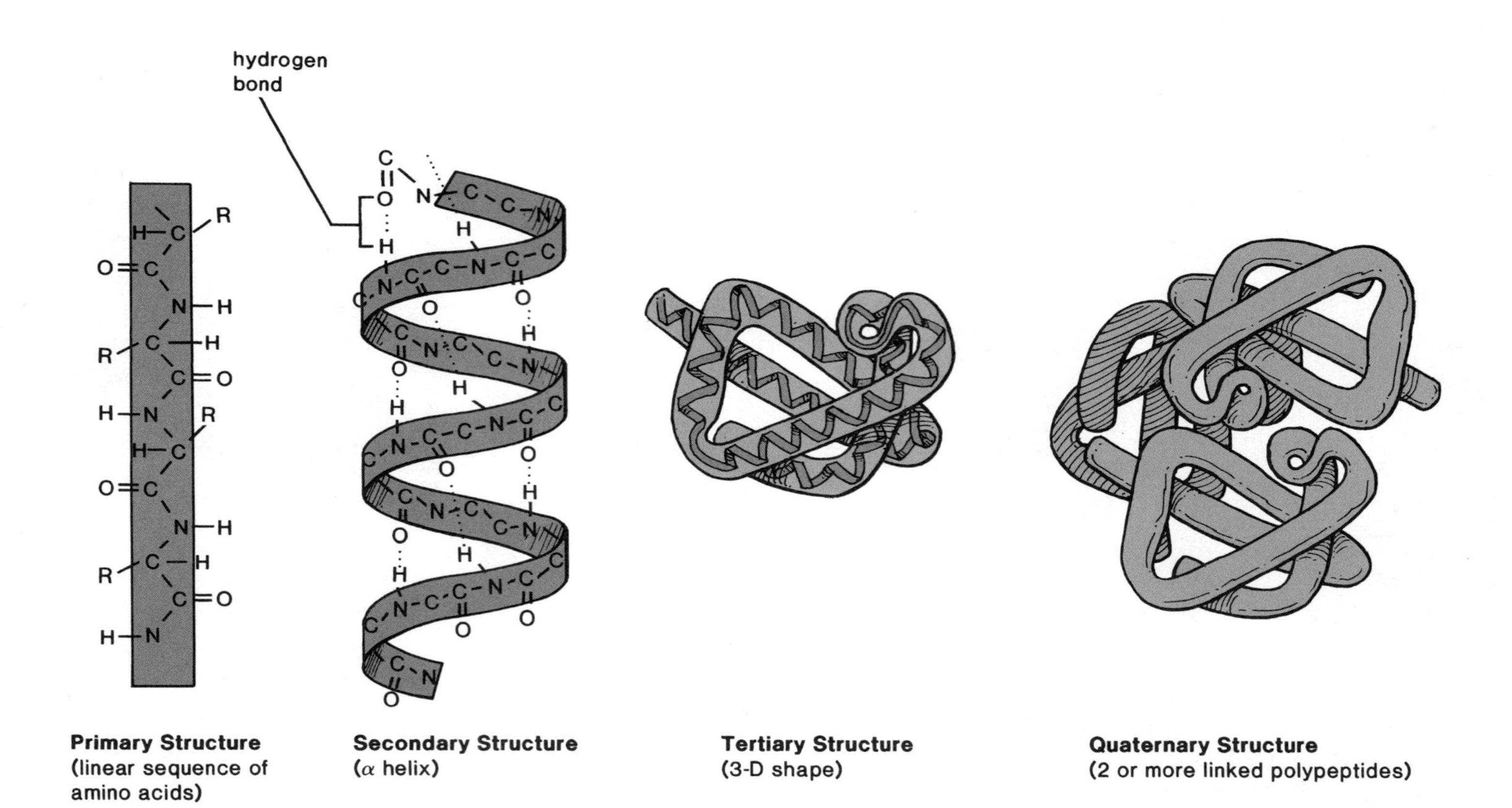
hydrogen
bond
Primary Structure
(linear sequence of
amino acids)
Secondary Structure
(α helix)
Tertiary Structure
(3-D shape)
Quaternary Structure
(2 or more linked polypeptides)

Levels of Protein Structure
Figure 2.12

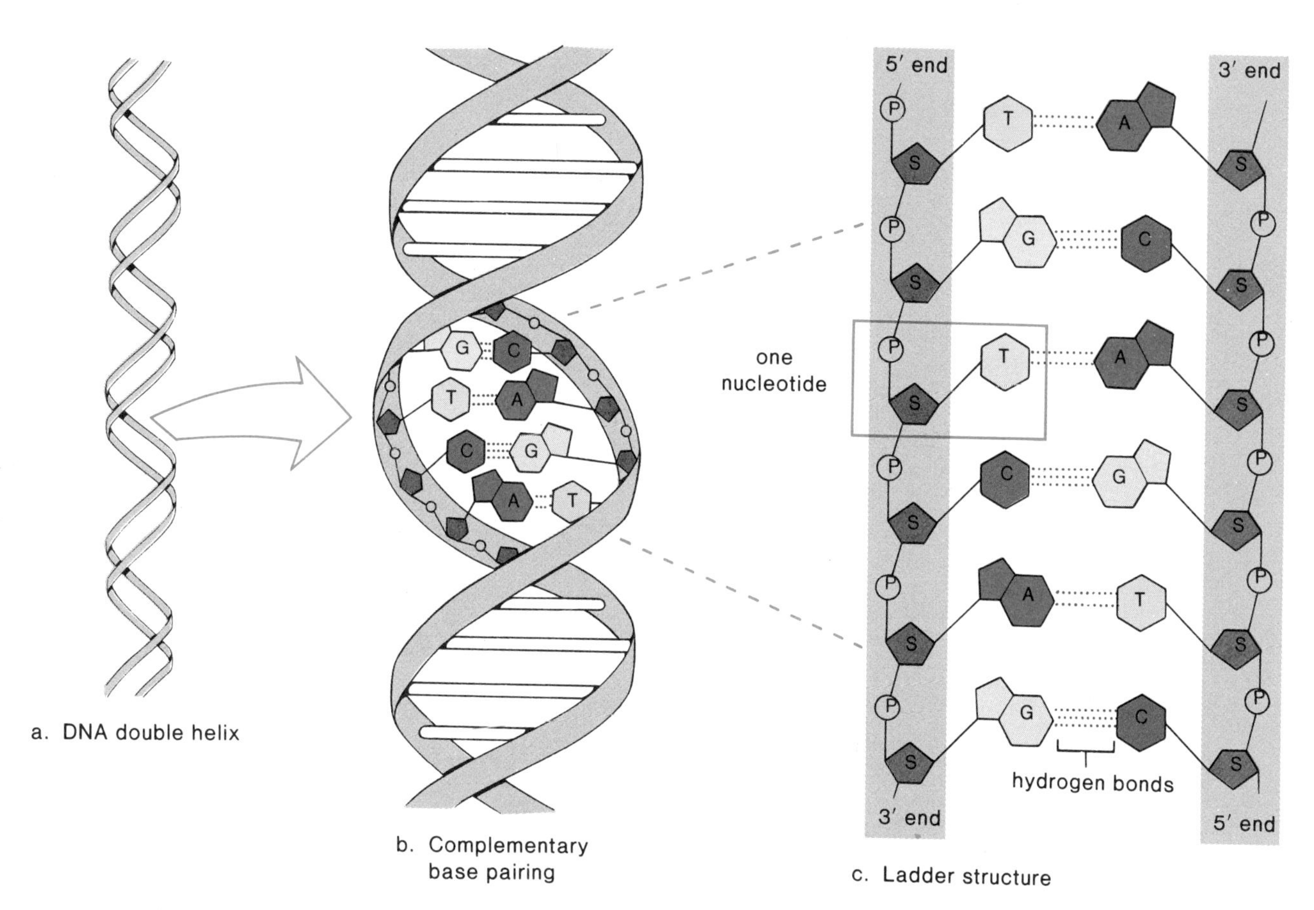
5′ end
3′ end
P
S
T
A
G
C
one
nucleotide
hydrogen bonds
3′ end
5′ end

a. DNA double helix

b. Complementary base pairing

c. Ladder structure

DNA Structure and Function
Figure 2.13

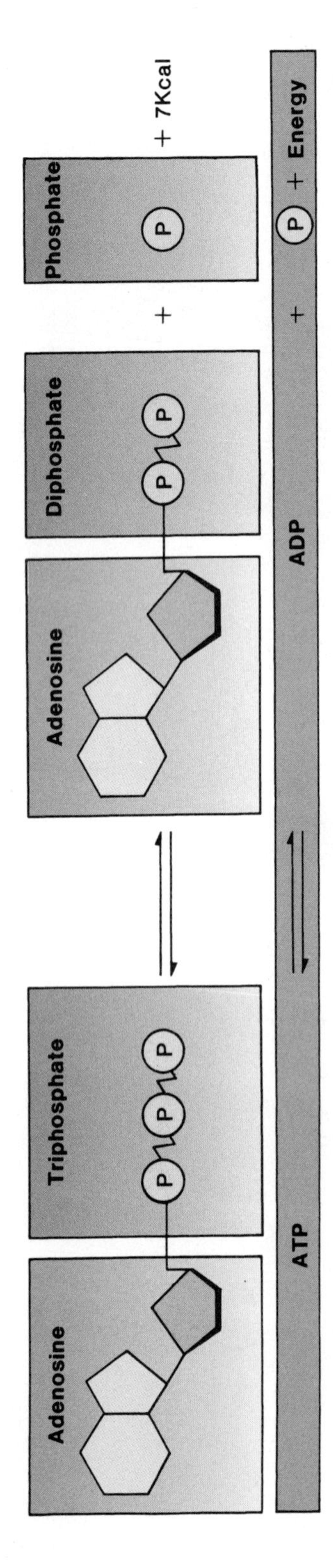
Adenosine
Triphosphate
P
P
P
Adenosine
Diphosphate
P
P
+
Phosphate
P
+ 7Kcal
ATP
ADP
+
P + Energy

ATP Reaction
Figure 2.14

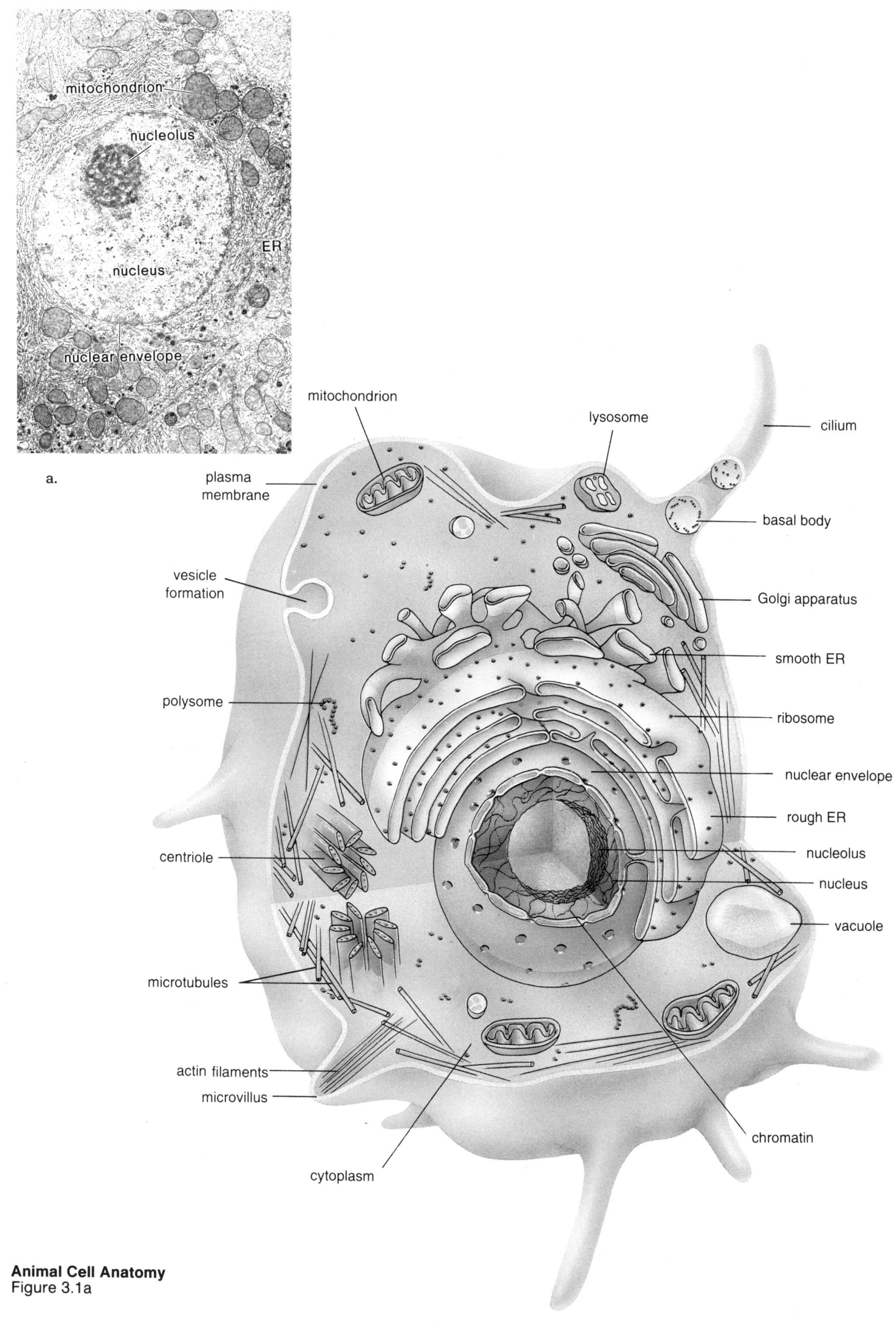
mitochondrion
nucleolus
ER
nucleus
nuclear envelope
a.
mitochondrion
lysosome
cilium
plasma membrane
basal body
vesicle formation
Golgi apparatus
smooth ER
polysome
ribosome
nuclear envelope
rough ER
centriole
nucleolus
nucleus
vacuole
microtubules
actin filaments
microvillus
chromatin
cytoplasm

Animal Cell Anatomy
Figure 3.1a

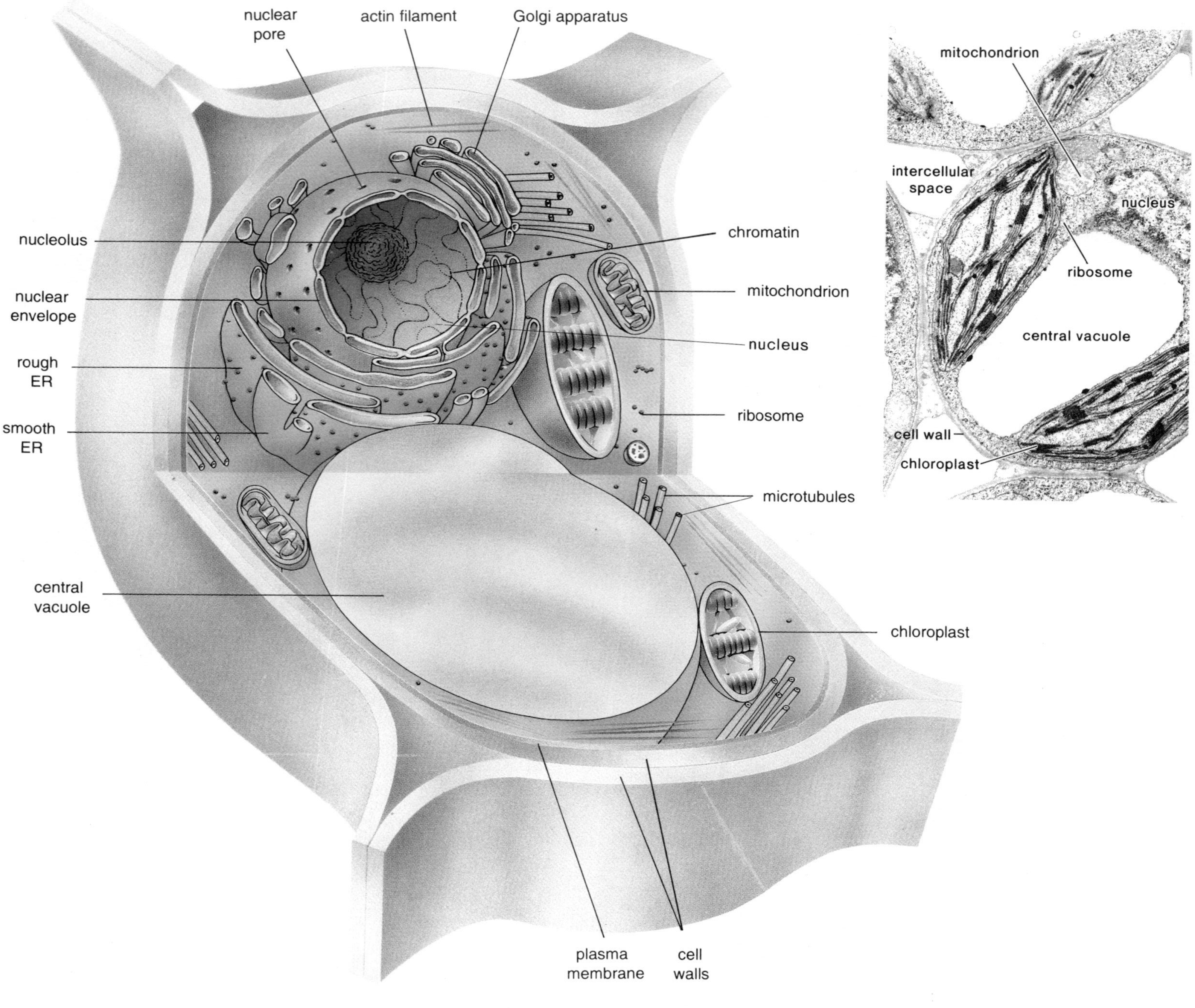

nuclear pore
actin filament
Golgi apparatus
nucleolus
nuclear envelope
rough ER
smooth ER
central vacuole
chromatin
mitochondrion
nucleus
ribosome
microtubules
chloroplast
plasma membrane
cell walls
mitochondrion
intercellular space
nucleus
ribosome
central vacuole
cell wall
chloroplast

Plant Cell Anatomy
Figure 3.1b

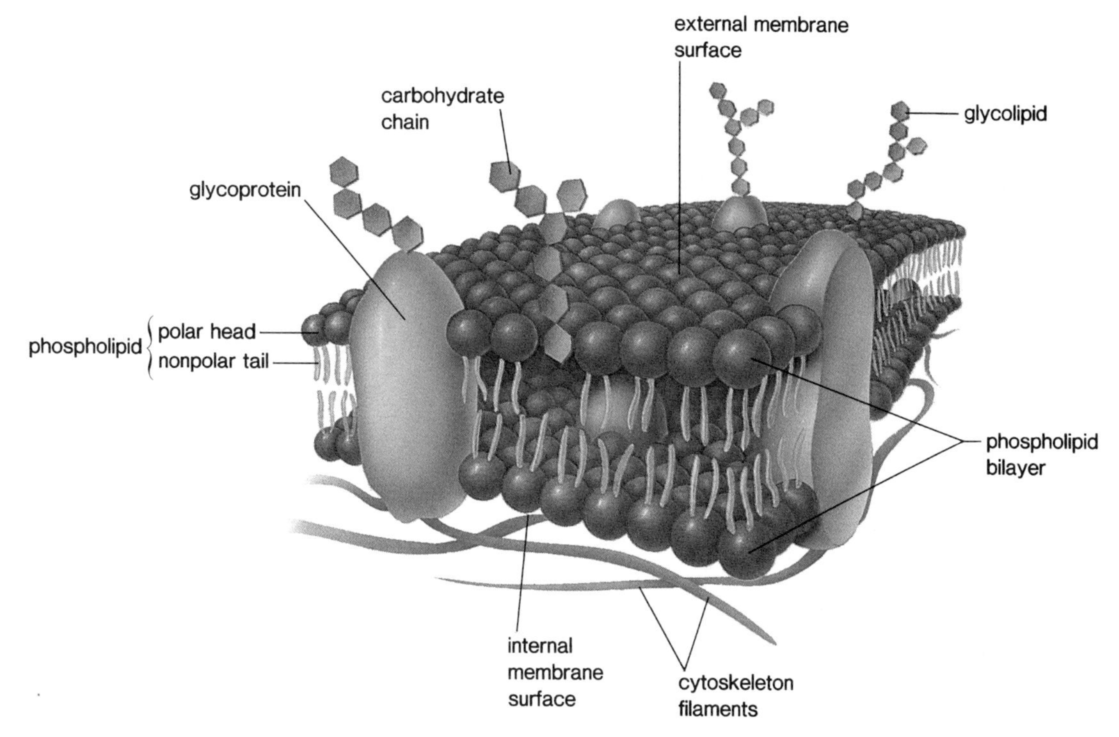
external membrane
surface
carbohydrate
chain
glycolipid
glycoprotein
phospholipid
polar head
nonpolar tail
phospholipid
bilayer
internal
membrane
surface
cytoskeleton
filaments

Plasma Membrane Structure
Figure 3.2

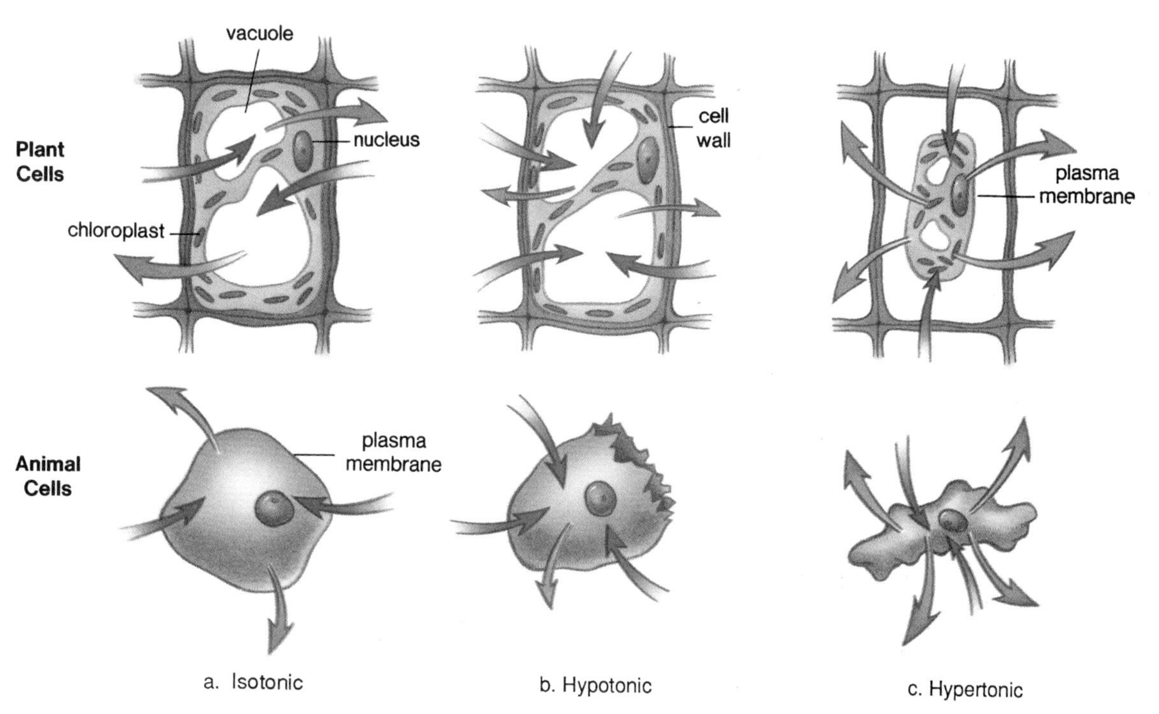
vacuole
nucleus
cell
wall
Plant
Cells
chloroplast
plasma
membrane
Animal
Cells
plasma
membrane
a. Isotonic
b. Hypotonic
c. Hypertonic

Effect of Tonicity in Plant and Animal Cells
Figure 3.3

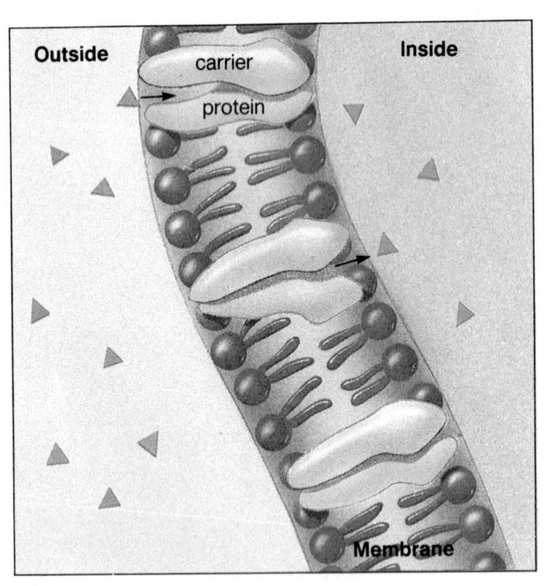
Outside
Inside
carrier
protein
Membrane
a. Facilitated transport

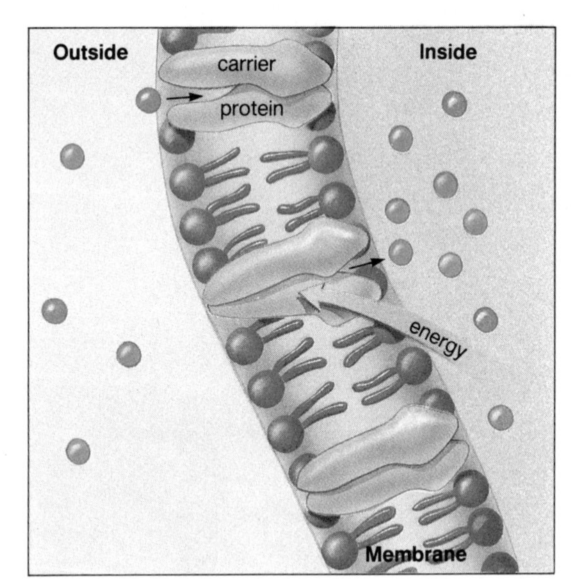
Outside
Inside
carrier
protein
energy
Membrane
b. Active transport

Transport Mechanisms
Figure 3.4

Endoplasmic Reticulum
Figure 3.5

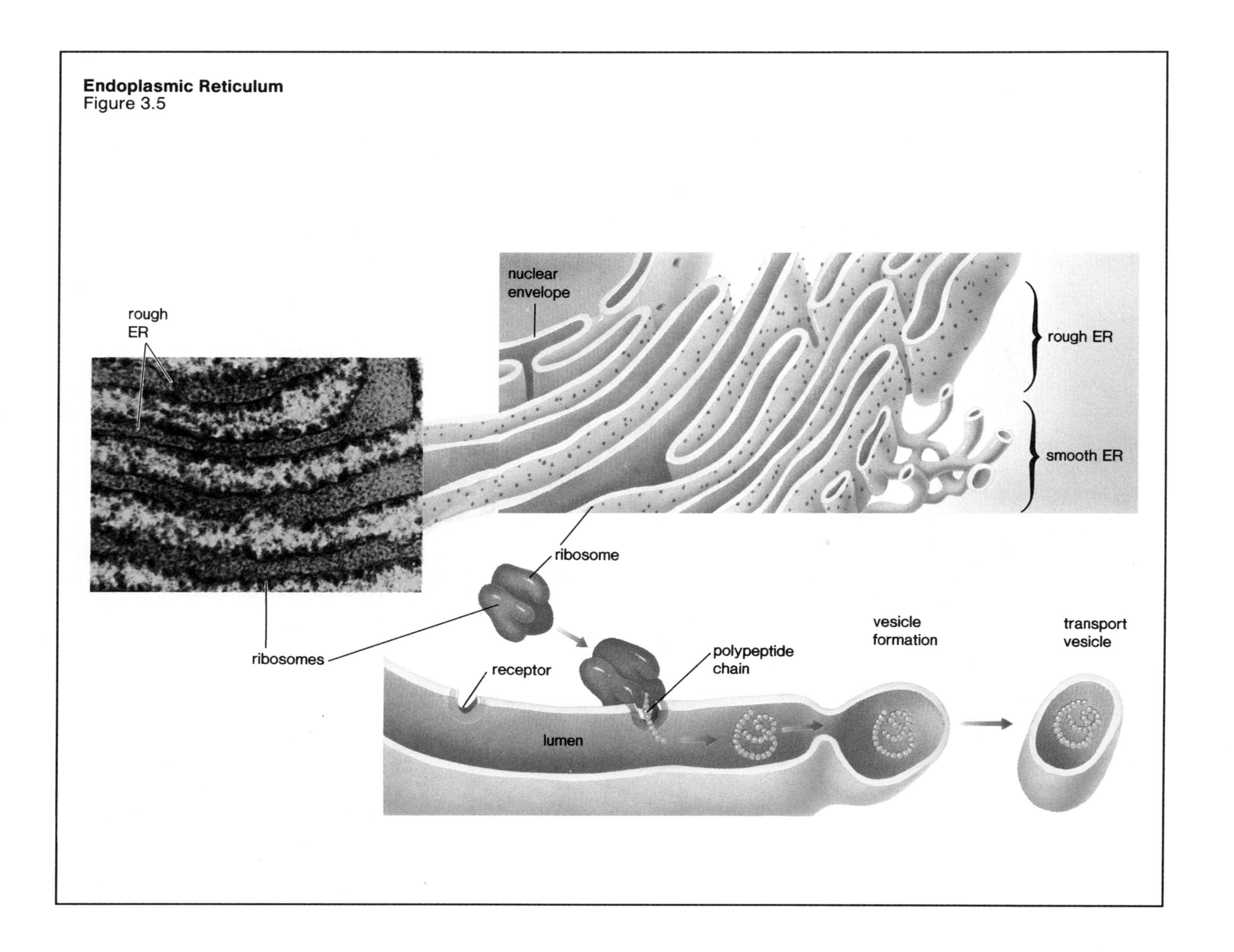

rough ER
nuclear envelope
rough ER
smooth ER
ribosome
ribosomes
receptor
lumen
polypeptide chain
vesicle formation
transport vesicle

Golgi Apparatus

Figure 3.6

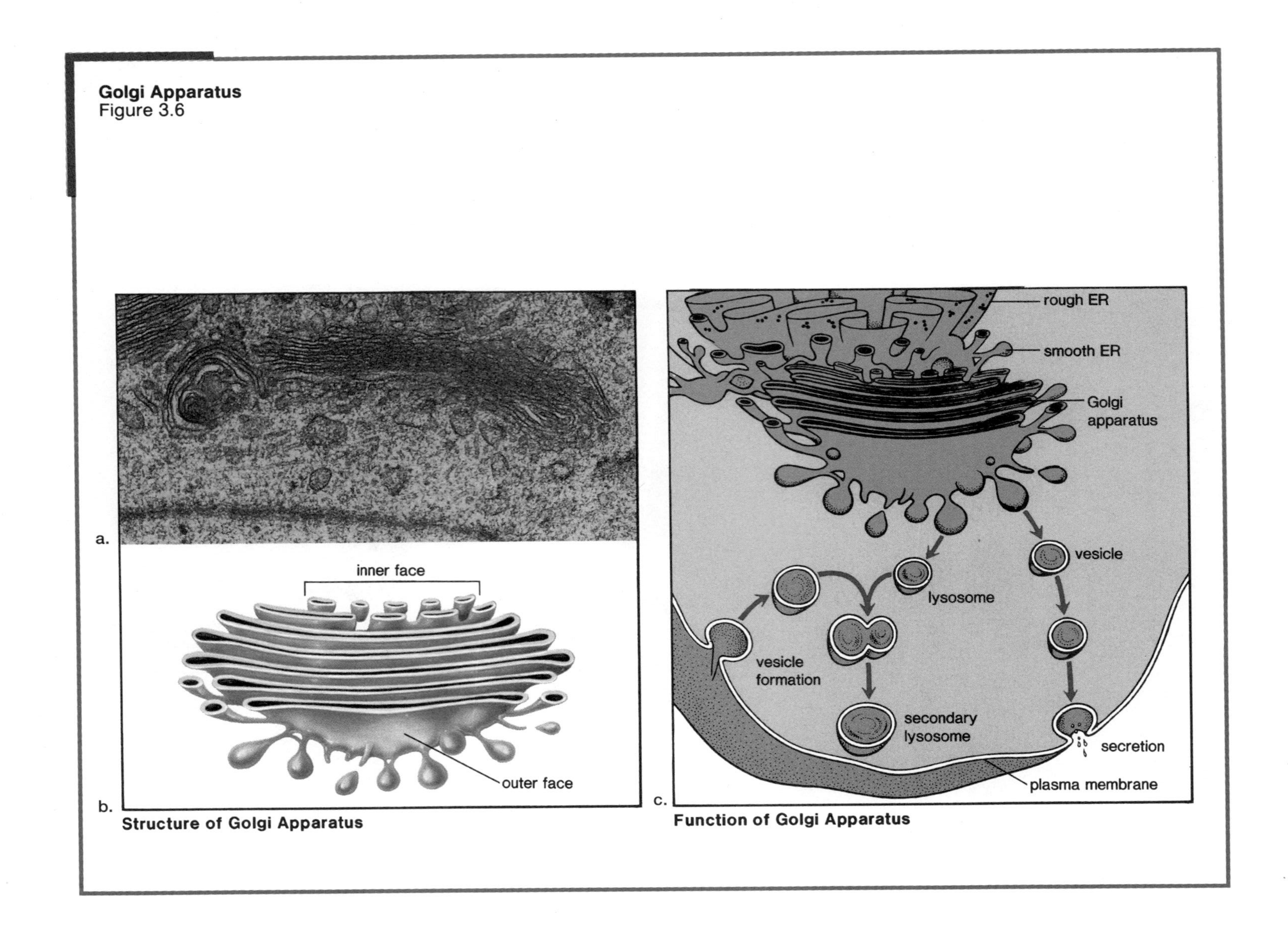
a.
b.
inner face
outer face
c.
rough ER
smooth ER
Golgi apparatus
vesicle
lysosome
vesicle formation
secondary lysosome
secretion
plasma membrane

b. Structure of Golgi Apparatus

c. Function of Golgi Apparatus

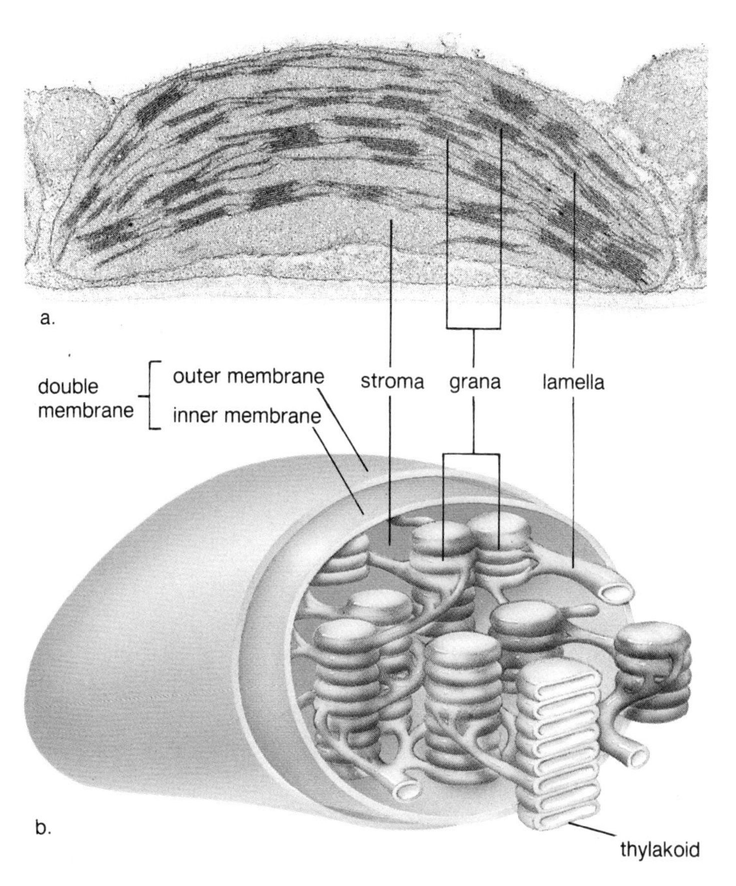
a.
double membrane
outer membrane
inner membrane
stroma
grana
lamella
b.
thylakoid

Chloroplast Structure
Figure 3.7

Prokaryotic Anatomy
Figure 3.12

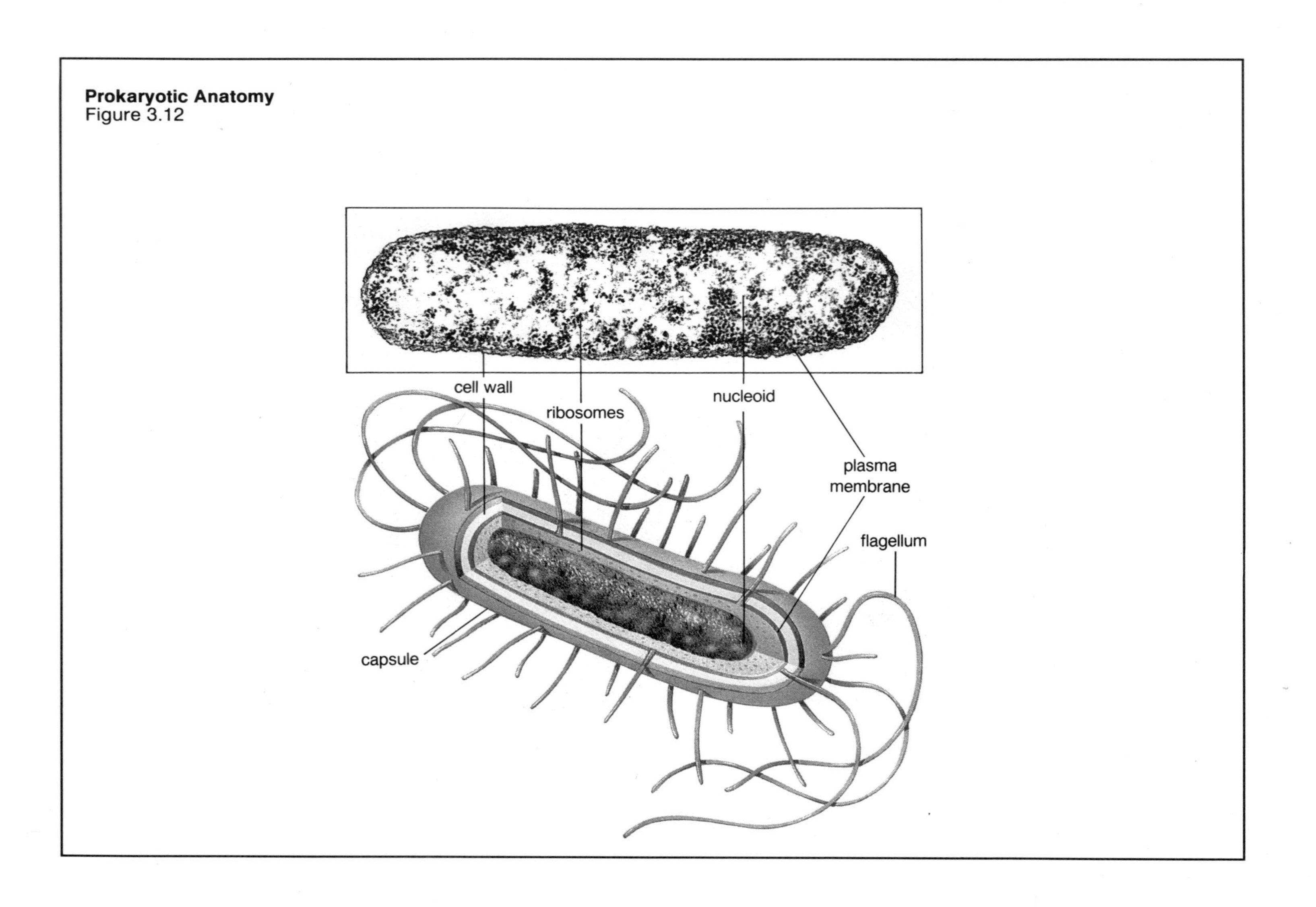
cell wall
ribosomes
nucleoid
plasma
membrane
flagellum
capsule

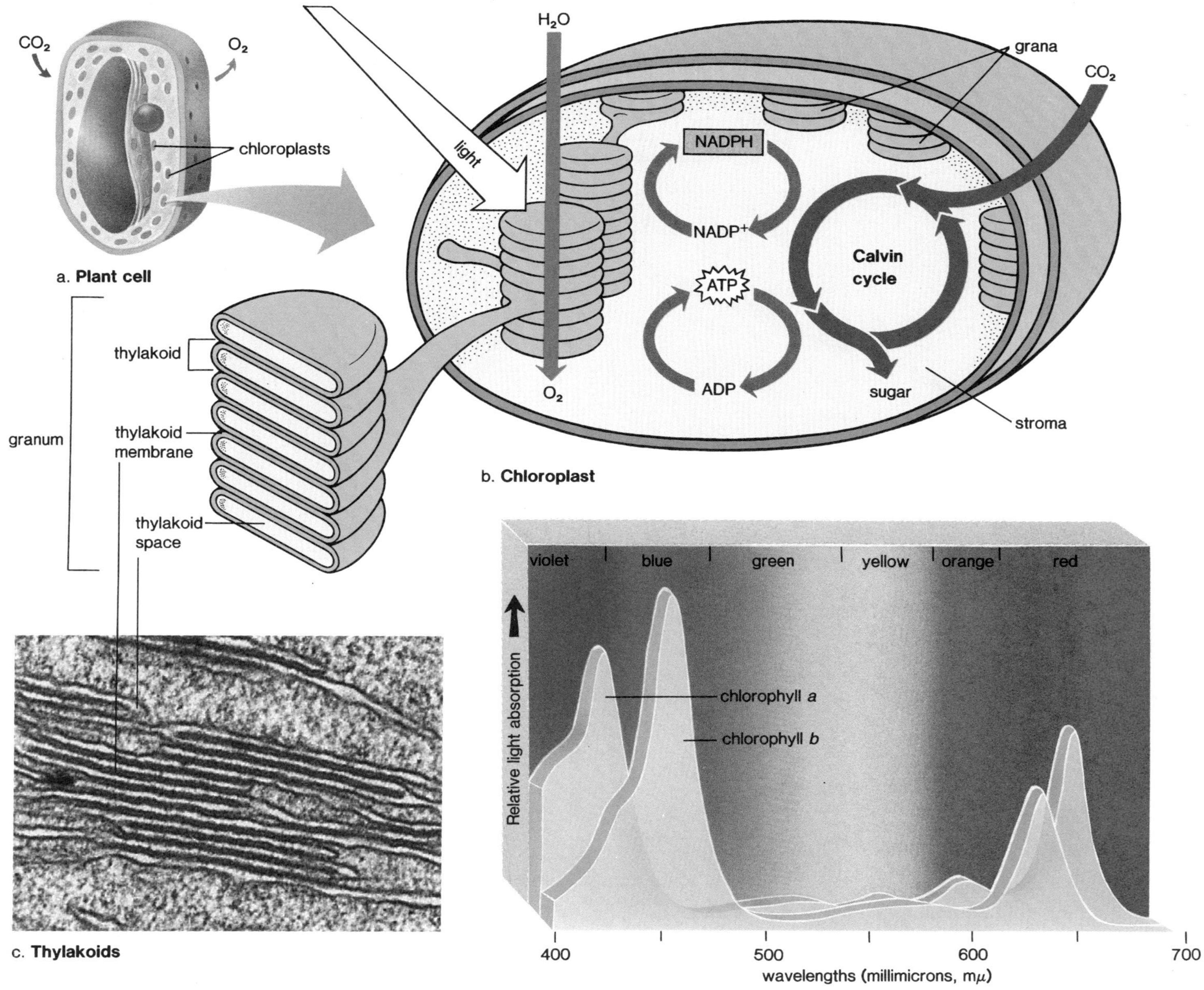

CO_2
O_2
chloroplasts
light
H_2O
O_2
grana
CO_2
NADPH
$NADP^+$
ATP
ADP
Calvin cycle
sugar
stroma
granum
thylakoid
thylakoid membrane
thylakoid space
violet
blue
green
yellow
orange
red
Relative light absorption
chlorophyll *a*
chlorophyll *b*
400
500
600
700
wavelengths (millimicrons, mμ)

a. **Plant cell**

b. **Chloroplast**

c. **Thylakoids**

d. **Absorption spectra for chlorophylls *a* and *b***

Chloroplast Structure and Function
Figure 4.4

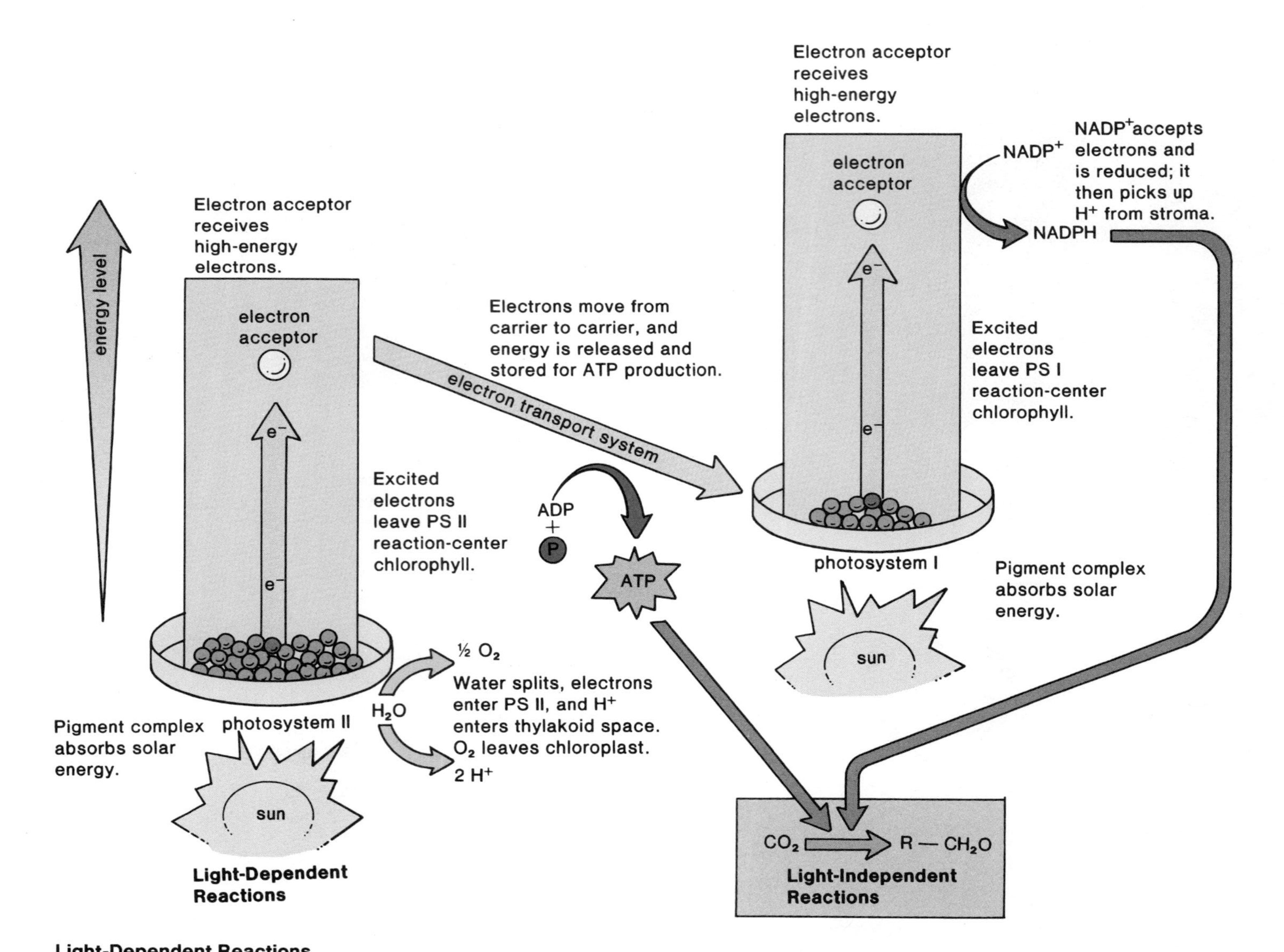
energy level
Electron acceptor receives high-energy electrons.
electron acceptor
e⁻
e⁻
Excited electrons leave PS II reaction-center chlorophyll.
Pigment complex absorbs solar energy.
photosystem II
sun
Light-Dependent Reactions
½ O_2
H_2O
Water splits, electrons enter PS II, and H^+ enters thylakoid space. O_2 leaves chloroplast.
2 H^+
Electrons move from carrier to carrier, and energy is released and stored for ATP production.
electron transport system
ADP
+
P
ATP
Electron acceptor receives high-energy electrons.
electron acceptor
e⁻
e⁻
Excited electrons leave PS I reaction-center chlorophyll.
photosystem I
sun
Pigment complex absorbs solar energy.
$NADP^+$
$NADP^+$ accepts electrons and is reduced; it then picks up H^+ from stroma.
NADPH
CO_2
R — CH_2O
Light-Independent Reactions

Light-Dependent Reactions
Figure 4.5

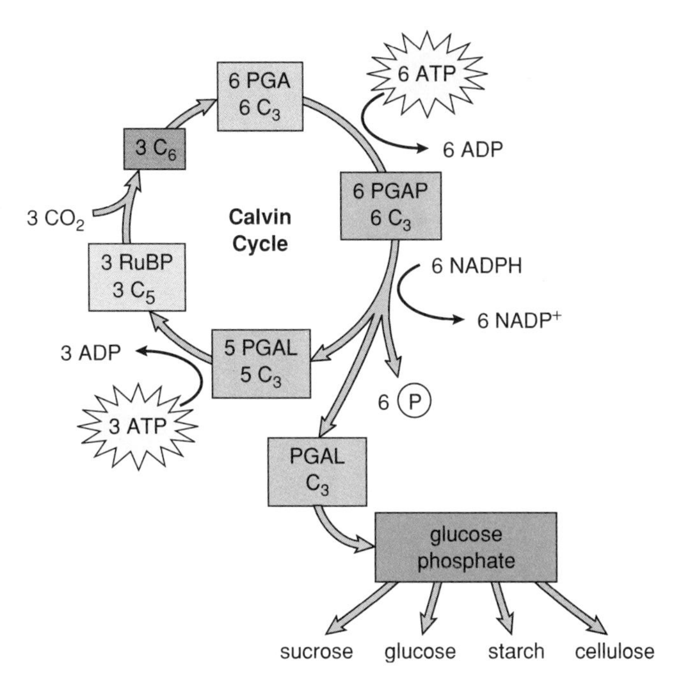
6 PGA
$6\ C_3$
6 ATP
6 ADP
$3\ C_6$
6 PGAP
$6\ C_3$
$3\ CO_2$
Calvin
Cycle
3 RuBP
$3\ C_5$
6 NADPH
$6\ NADP^+$
3 ADP
5 PGAL
$5\ C_3$
6 P
3 ATP
PGAL
C_3
glucose
phosphate
sucrose
glucose
starch
cellulose

The Calvin Cycle
Figure 4.6

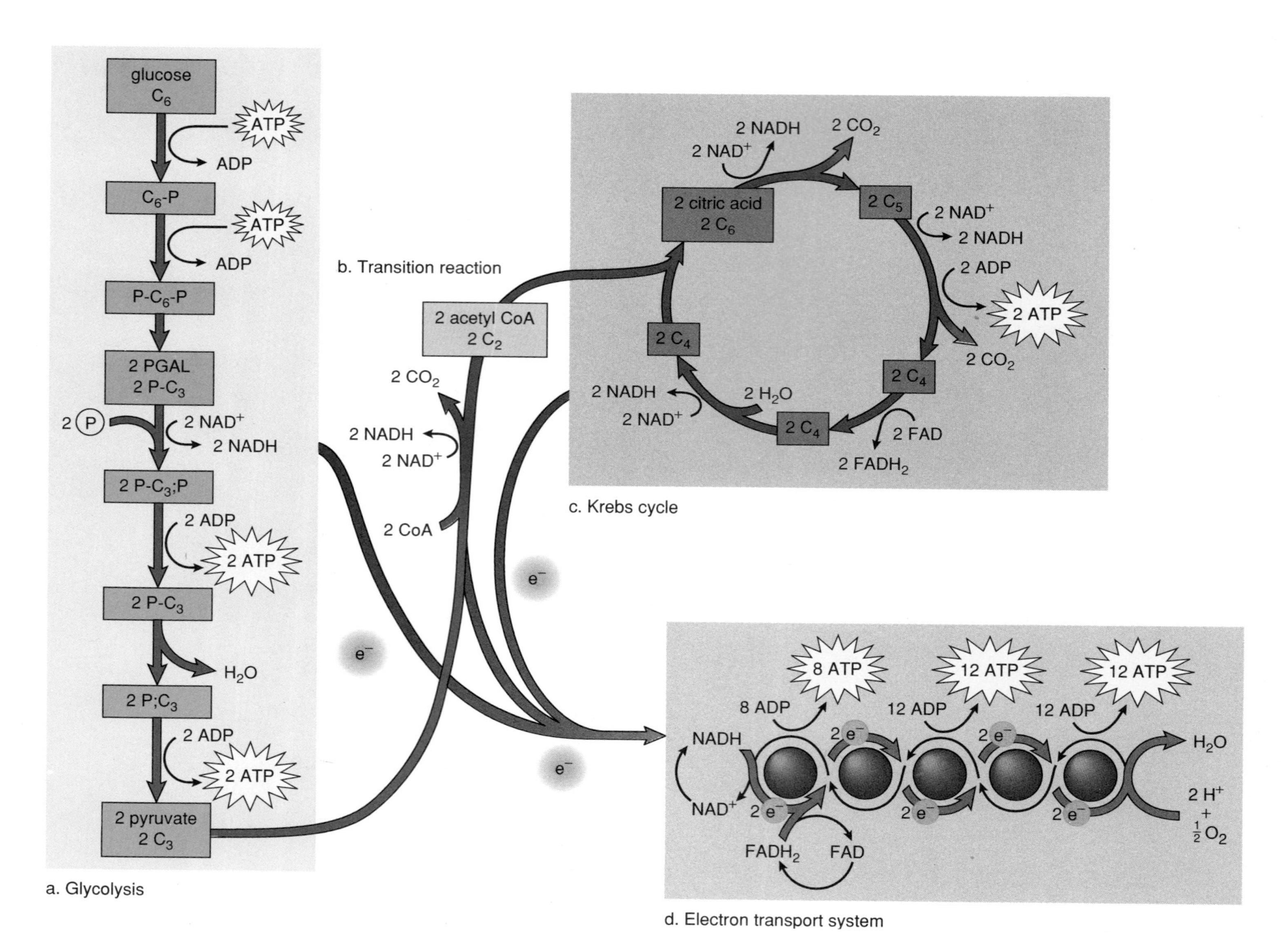
glucose
C_6
ATP
ADP
C_6-P
ATP
ADP
P-C_6-P
2 PGAL
2 P-C_3
2 (P)
2 NAD^+
2 NADH
2 P-C_3;P
2 ADP
2 ATP
2 P-C_3
H_2O
2 P;C_3
2 ADP
2 ATP
2 pyruvate
2 C_3
a. Glycolysis
b. Transition reaction
2 acetyl CoA
2 C_2
2 CO_2
2 NADH
2 NAD^+
2 CoA
e^-
2 NADH
2 CO_2
2 NAD^+
2 citric acid
2 C_6
2 C_5
2 NAD^+
2 NADH
2 ADP
2 ATP
2 CO_2
2 C_4
2 FAD
2 $FADH_2$
2 C_4
2 H_2O
2 C_4
2 NADH
2 NAD^+
c. Krebs cycle
e^-
e^-
8 ATP
12 ATP
12 ATP
8 ADP
12 ADP
12 ADP
NADH
NAD^+
2 e^-
2 e^-
2 e^-
2 e^-
2 e^-
$FADH_2$
FAD
H_2O
2 H^+
+
$\frac{1}{2} O_2$
d. Electron transport system

Aerobic Respiration
Figure 4.7

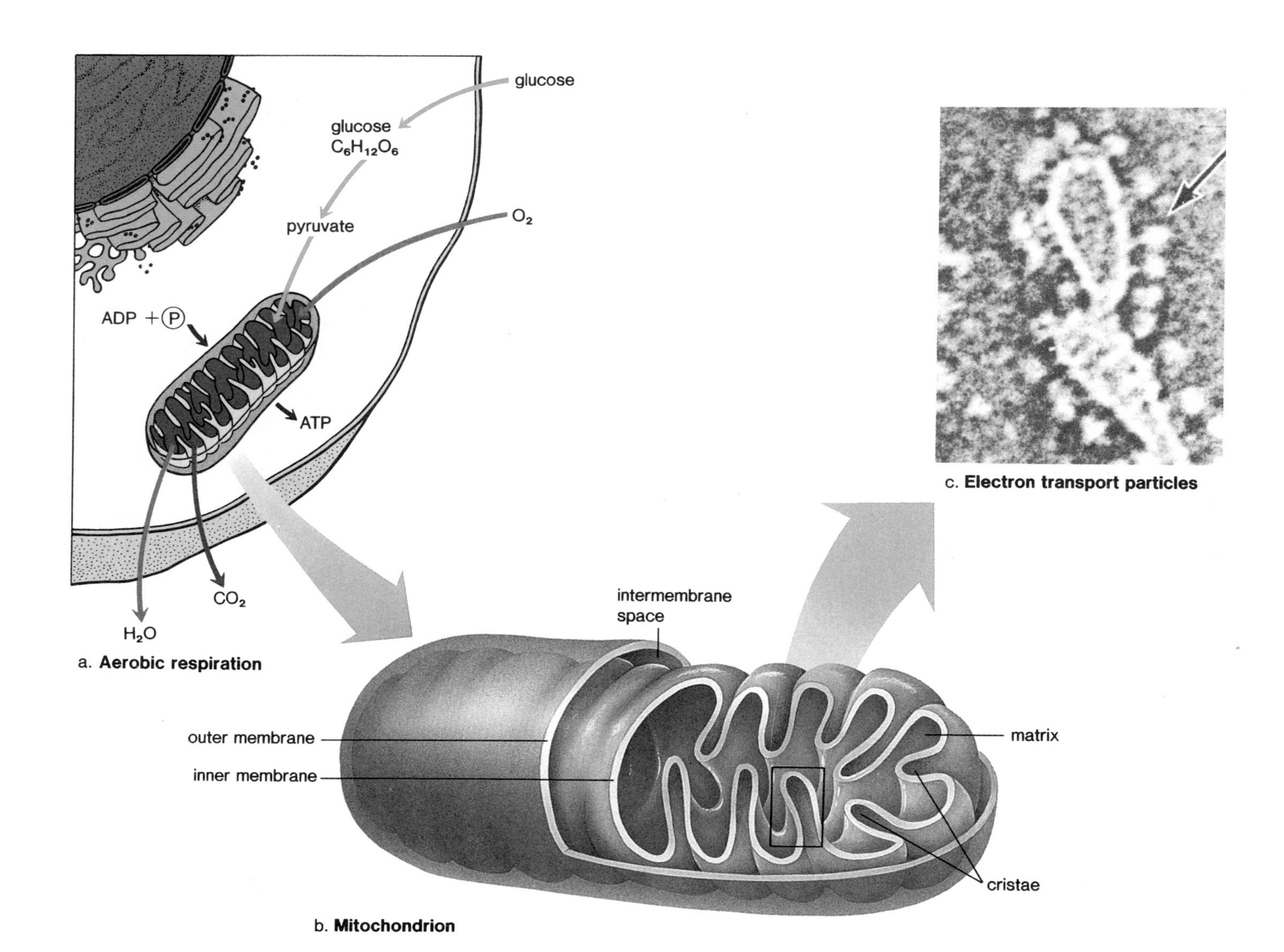
glucose
glucose
$C_6H_{12}O_6$
pyruvate
O_2
ADP + (P)
ATP
CO_2
H_2O
a. Aerobic respiration
intermembrane
space
outer membrane
inner membrane
matrix
cristae
b. Mitochondrion
c. Electron transport particles

Mitochondrion Structure and Function
Figure 4.8

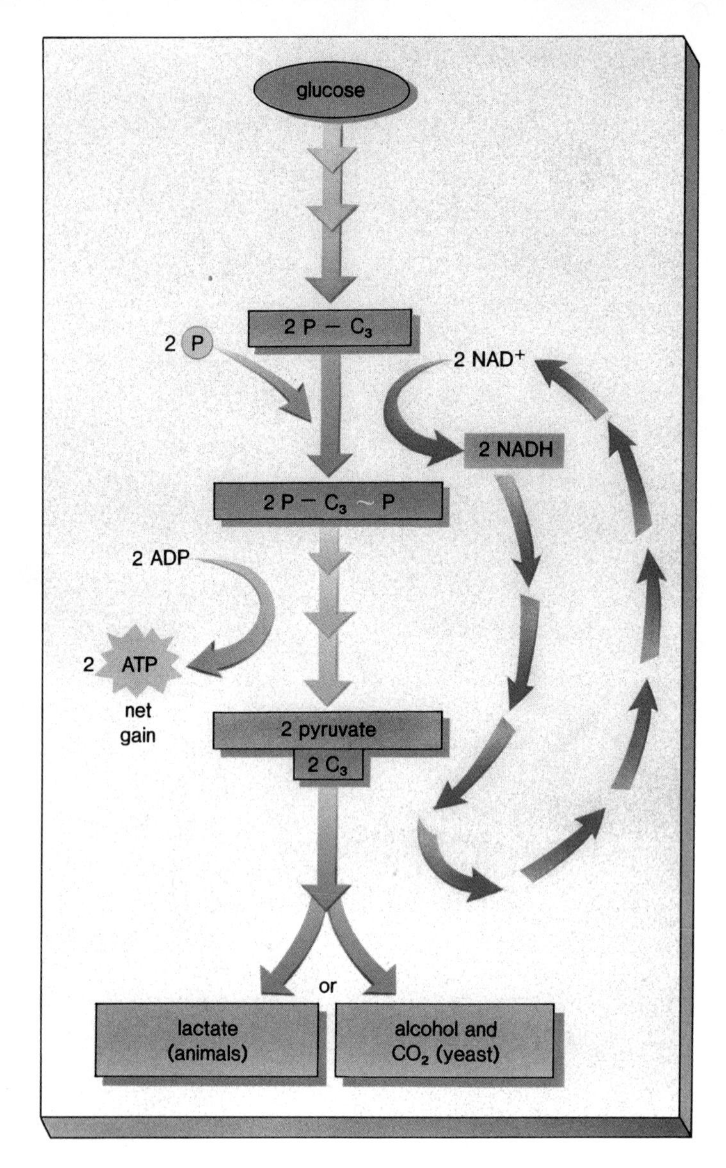
glucose
2 P − C_3
2 P
2 NAD^+
2 NADH
2 P − C_3 ~ P
2 ADP
2 ATP
net
gain
2 pyruvate
2 C_3
or
lactate
(animals)
alcohol and
CO_2 (yeast)

Fermentation
Figure 4.9

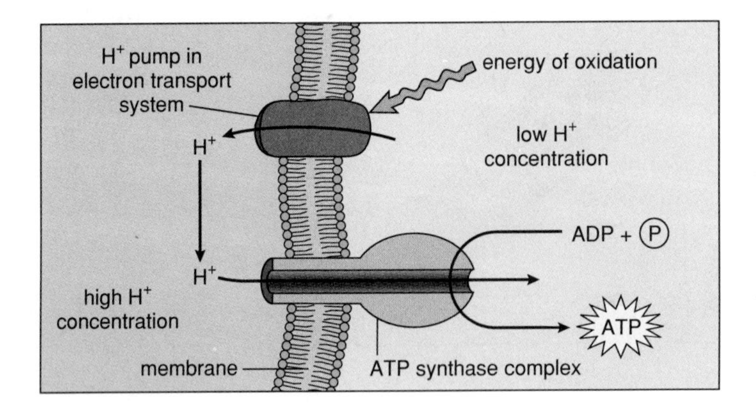
H^+ pump in
electron transport
system
energy of oxidation
H^+
low H^+
concentration
ADP + P
H^+
high H^+
concentration
ATP
membrane
ATP synthase complex

Chemiosmosis
Figure 4.10

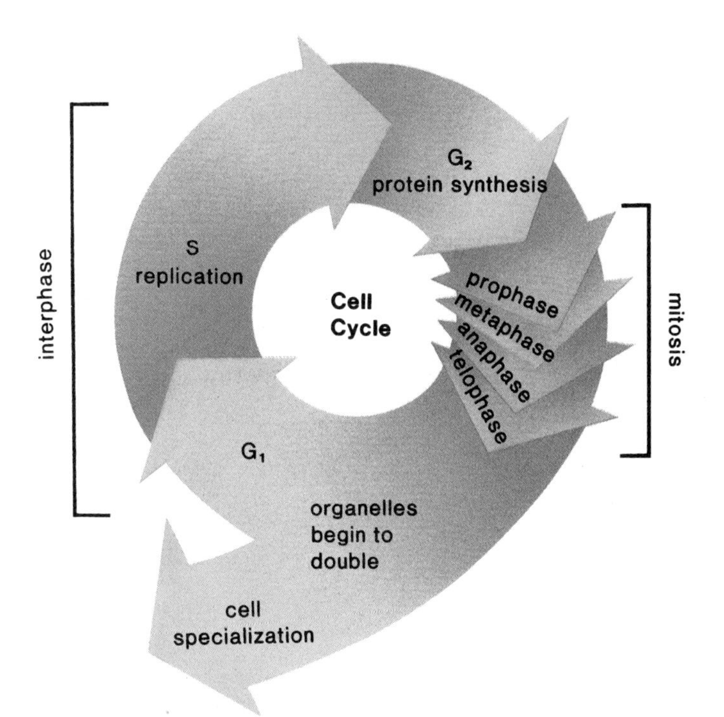
interphase
S
replication
G_2
protein synthesis
Cell
Cycle
prophase
metaphase
anaphase
telophase
mitosis
G_1
organelles
begin to
double
cell
specialization

The Cell Cycle
Figure 5.1

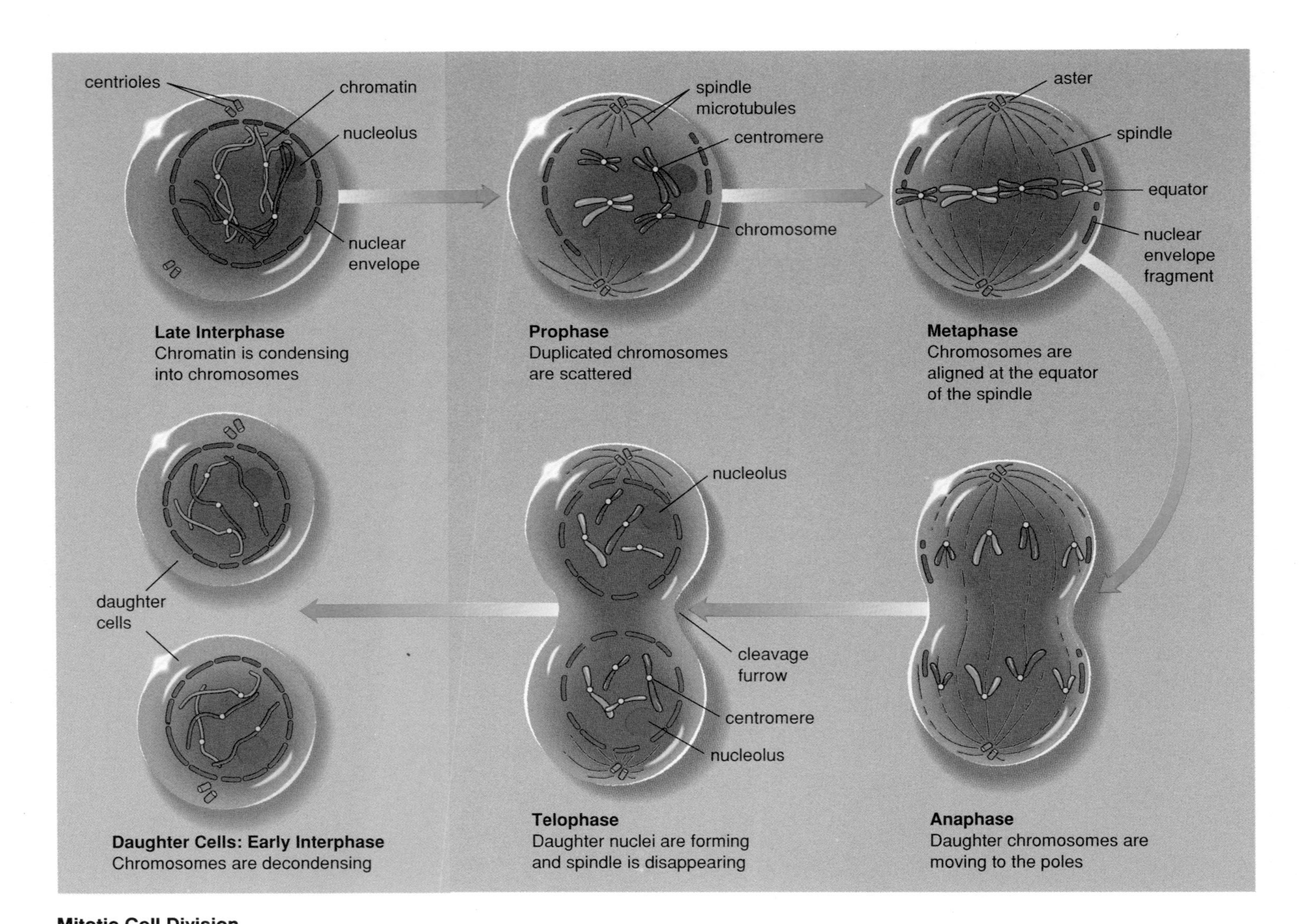
centrioles
chromatin
nucleolus
nuclear envelope
Late Interphase
Chromatin is condensing into chromosomes
spindle microtubules
centromere
chromosome
Prophase
Duplicated chromosomes are scattered
aster
spindle
equator
nuclear envelope fragment
Metaphase
Chromosomes are aligned at the equator of the spindle
daughter cells
Daughter Cells: Early Interphase
Chromosomes are decondensing
nucleolus
cleavage furrow
centromere
nucleolus
Telophase
Daughter nuclei are forming and spindle is disappearing
Anaphase
Daughter chromosomes are moving to the poles

Mitotic Cell Division
Figure 5.3

Telophase and Plant Cell Cytokinesis
Figure 5.5

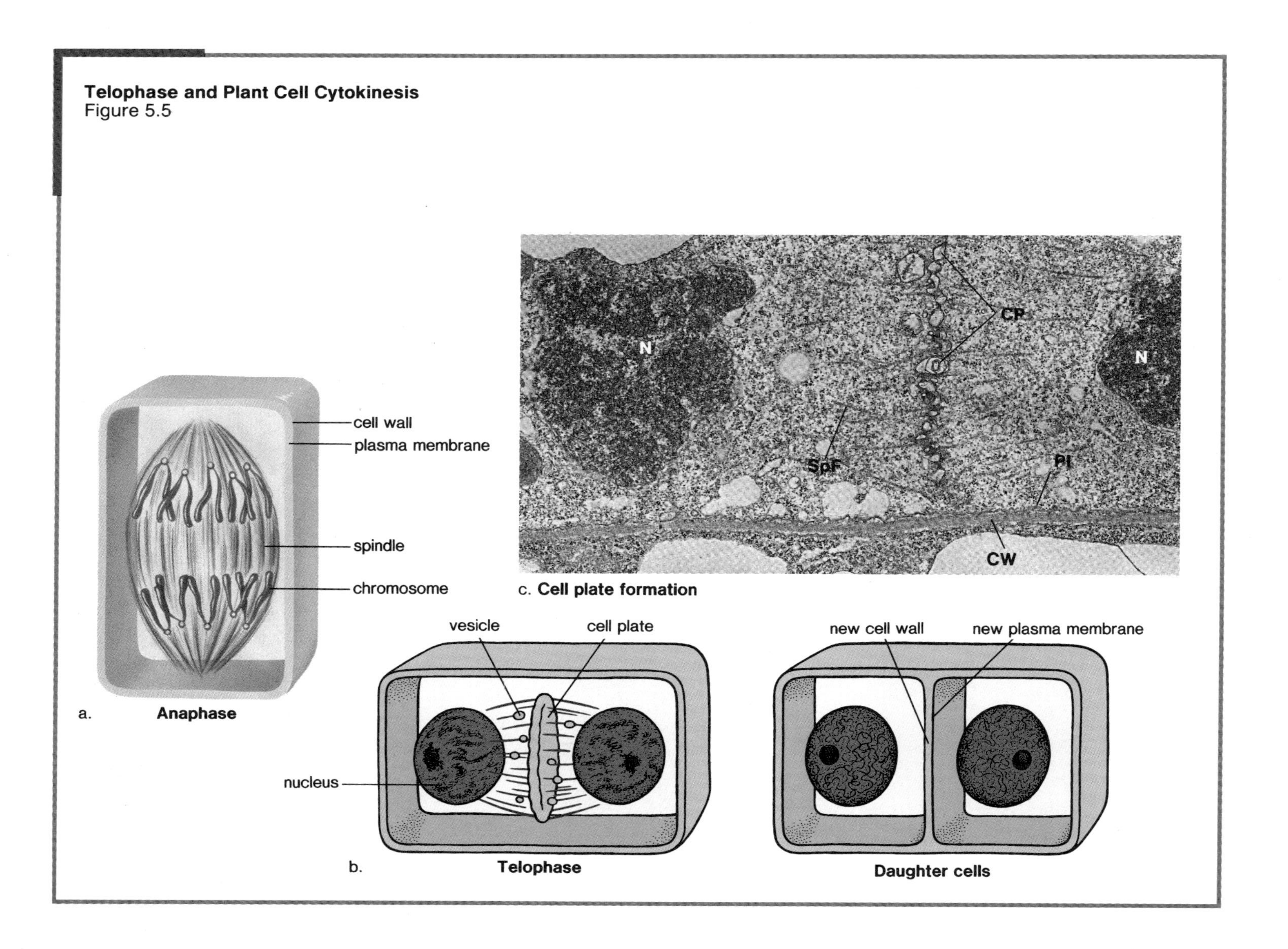
cell wall
plasma membrane
spindle
chromosome
a.
Anaphase
N
CP
N
SpF
Pl
CW
c. Cell plate formation
vesicle
cell plate
nucleus
b.
Telophase
new cell wall
new plasma membrane
Daughter cells

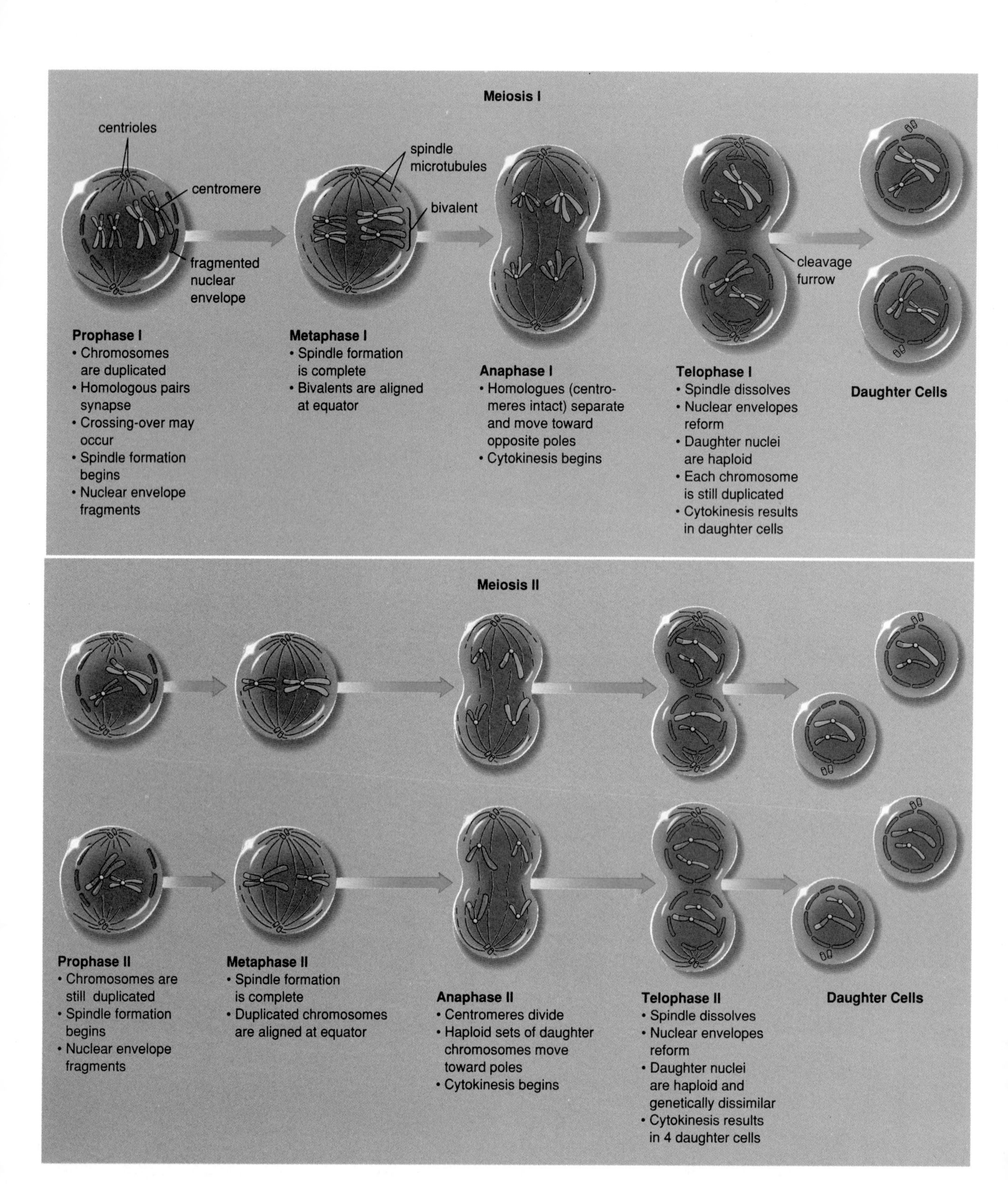
Meiosis I
centrioles
centromere
spindle
microtubules
bivalent
fragmented
nuclear
envelope
cleavage
furrow
Prophase I
• Chromosomes are duplicated
• Homologous pairs synapse
• Crossing-over may occur
• Spindle formation begins
• Nuclear envelope fragments
Metaphase I
• Spindle formation is complete
• Bivalents are aligned at equator
Anaphase I
• Homologues (centromeres intact) separate and move toward opposite poles
• Cytokinesis begins
Telophase I
• Spindle dissolves
• Nuclear envelopes reform
• Daughter nuclei are haploid
• Each chromosome is still duplicated
• Cytokinesis results in daughter cells
Daughter Cells
Meiosis II
Prophase II
• Chromosomes are still duplicated
• Spindle formation begins
• Nuclear envelope fragments
Metaphase II
• Spindle formation is complete
• Duplicated chromosomes are aligned at equator
Anaphase II
• Centromeres divide
• Haploid sets of daughter chromosomes move toward poles
• Cytokinesis begins
Telophase II
• Spindle dissolves
• Nuclear envelopes reform
• Daughter nuclei are haploid and genetically dissimilar
• Cytokinesis results in 4 daughter cells
Daughter Cells

Meiosis I and Meiosis II
Figure 5.8

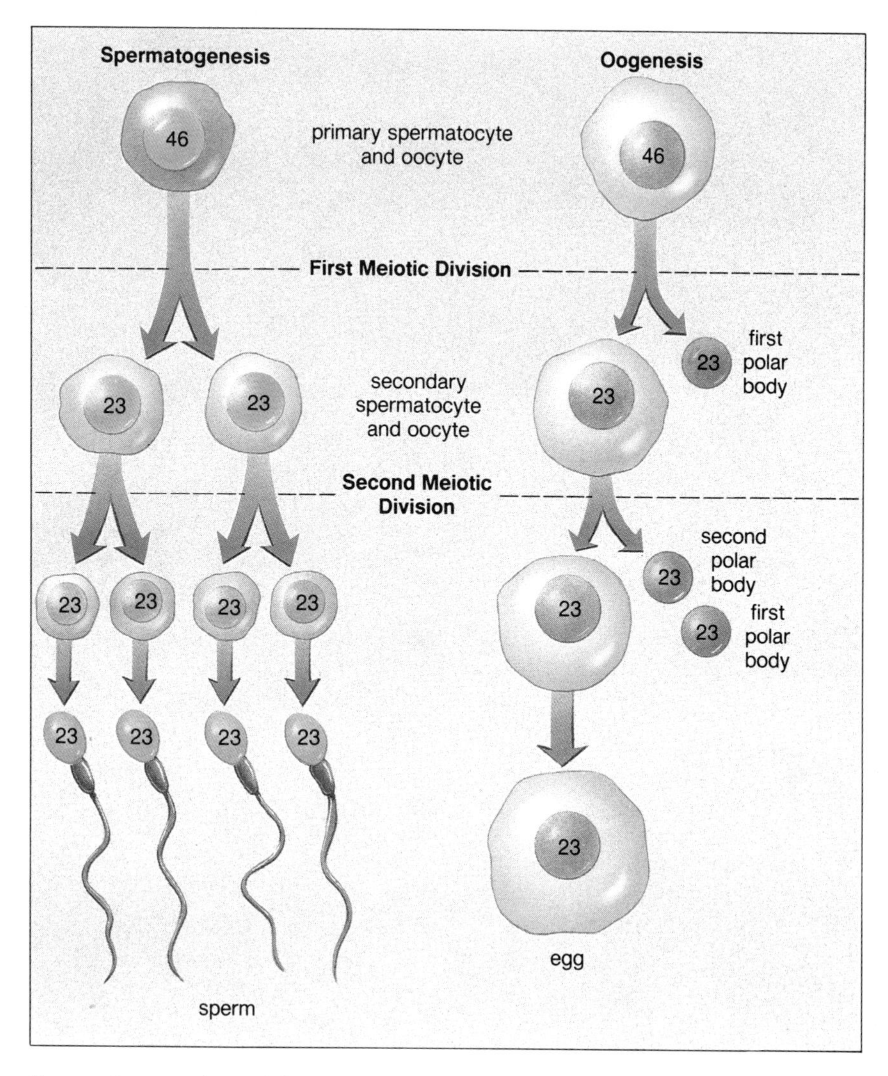
Spermatogenesis
Oogenesis
46
primary spermatocyte
and oocyte
46
First Meiotic Division
first
polar
body
23
23
23
secondary
spermatocyte
and oocyte
23
Second Meiotic
Division
second
polar
body
23
23
23
23
23
23
first
polar
body
23
23
23
23
23
23
sperm
egg

Spermatogenesis and Oogenesis
Figure 5.9

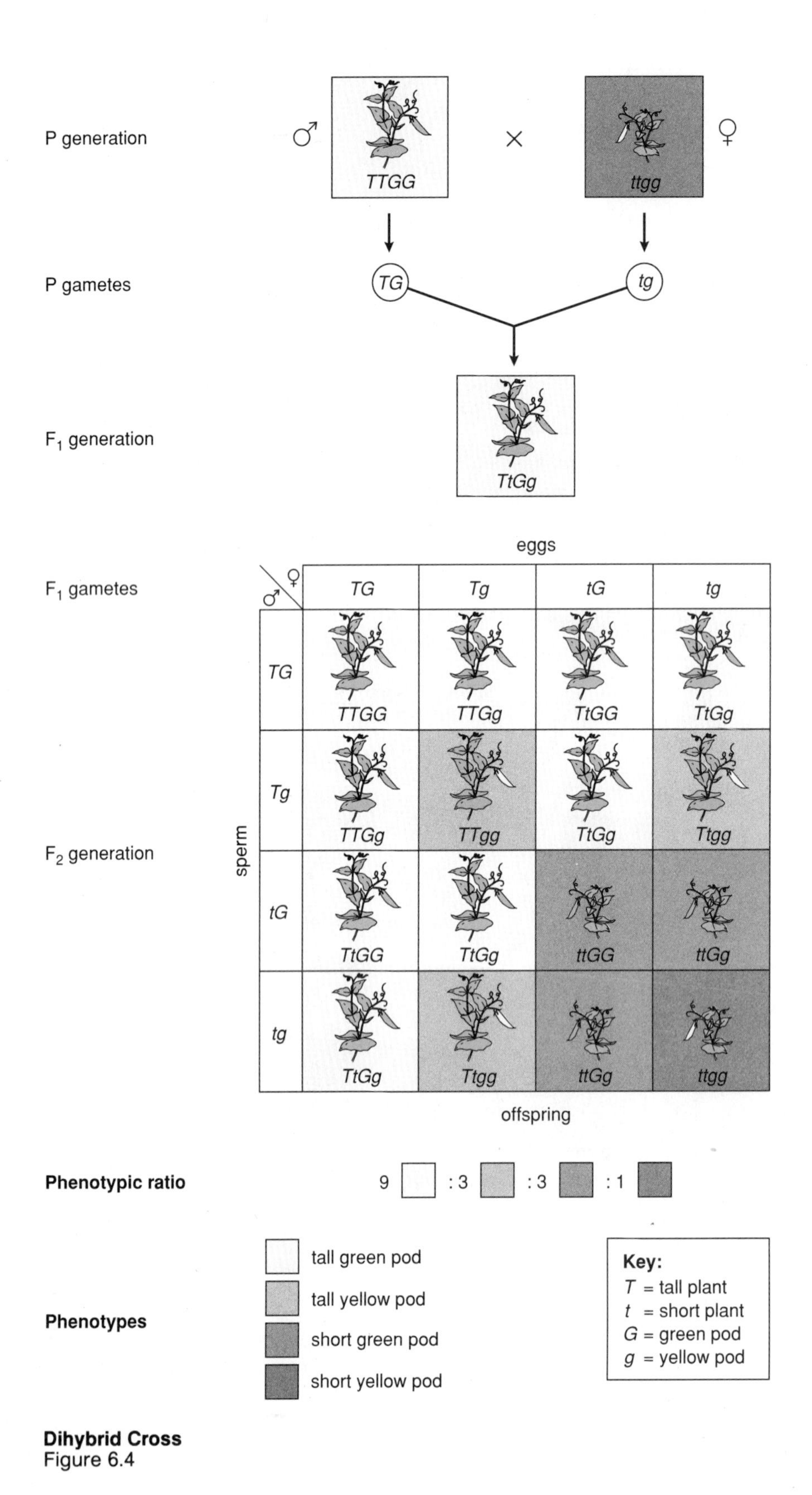
P generation
♂
TTGG
×
ttgg
♀
P gametes
TG
tg
F_1 generation
TtGg
eggs
F_1 gametes
♀
♂
TG
Tg
tG
tg
TG
TTGG
TTGg
TtGG
TtGg
Tg
TTGg
TTgg
TtGg
Ttgg
F_2 generation
sperm
tG
TtGG
TtGg
ttGG
ttGg
tg
TtGg
Ttgg
ttGg
ttgg
offspring
Phenotypic ratio
9 : 3 : 3 : 1
Phenotypes
tall green pod
tall yellow pod
short green pod
short yellow pod
Key:
T = tall plant
t = short plant
G = green pod
g = yellow pod

Dihybrid Cross
Figure 6.4

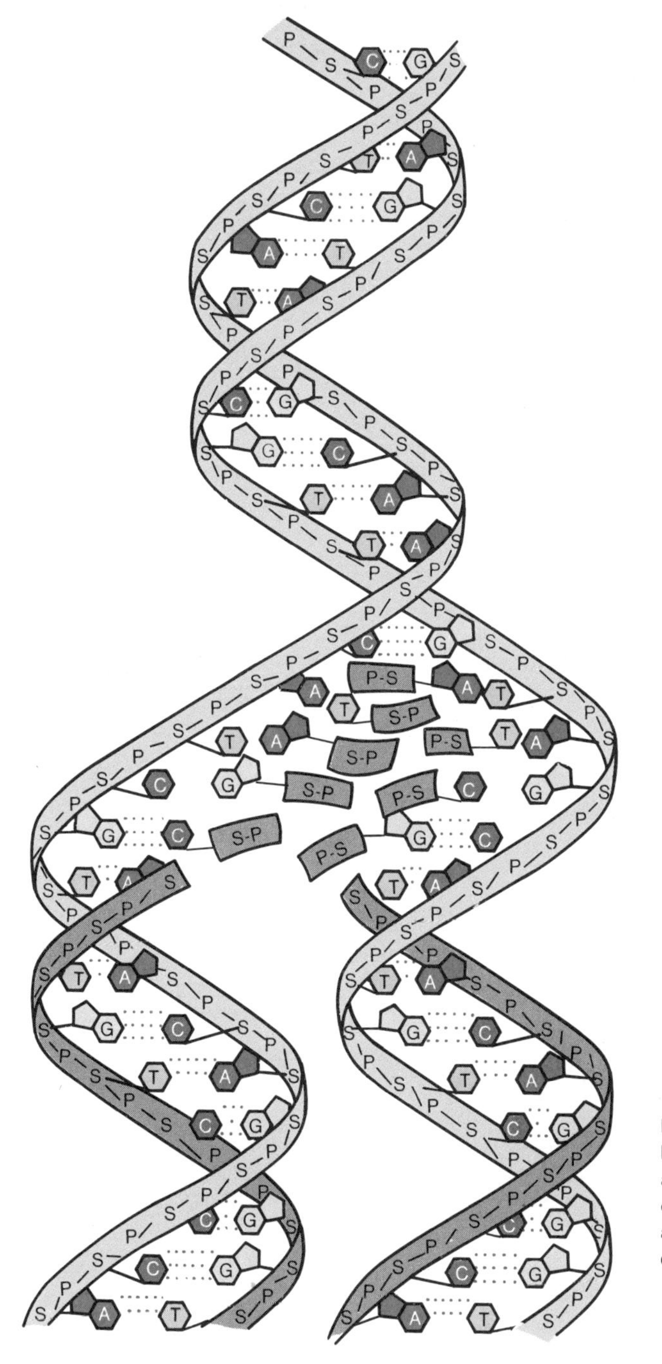
Region of parental DNA helix. (Both backbones are light.)
Region of replication (simplified). Parental DNA is unwound and unzipped. New nucleotides are pairing with those in parental strands.
Region of completed replication. Each double helix is composed of an old parental strand (light) and a new daughter strand (dark). Notice that each double helix is exactly like the other and also like the original parental double helix.

DNA Replication
Figure 7.2

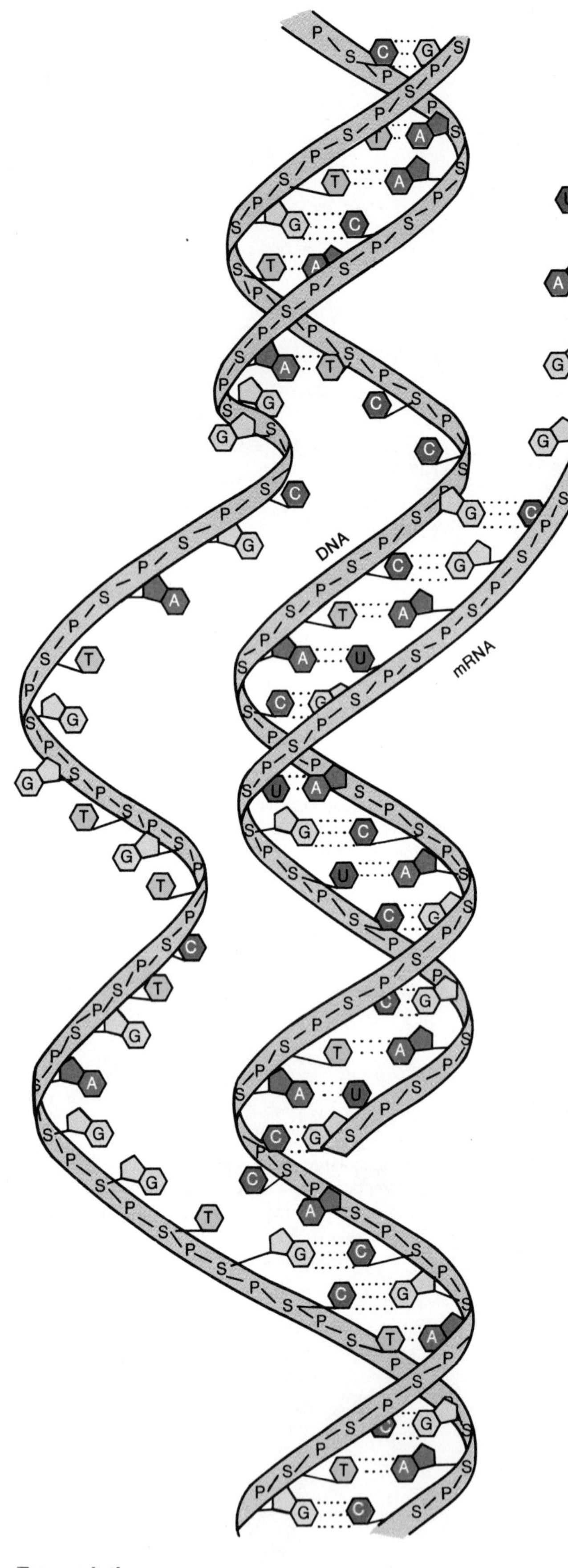
DNA
mRNA

3. This mRNA transcript is ready to move into the cytoplasm.

2. Transcription is going on here—the nucleotides of mRNA are joined in an order complementary to a strand of DNA.

1. One portion of DNA—a particular gene or genes—is transcribed at a time.

Transcription
Figure 7.4

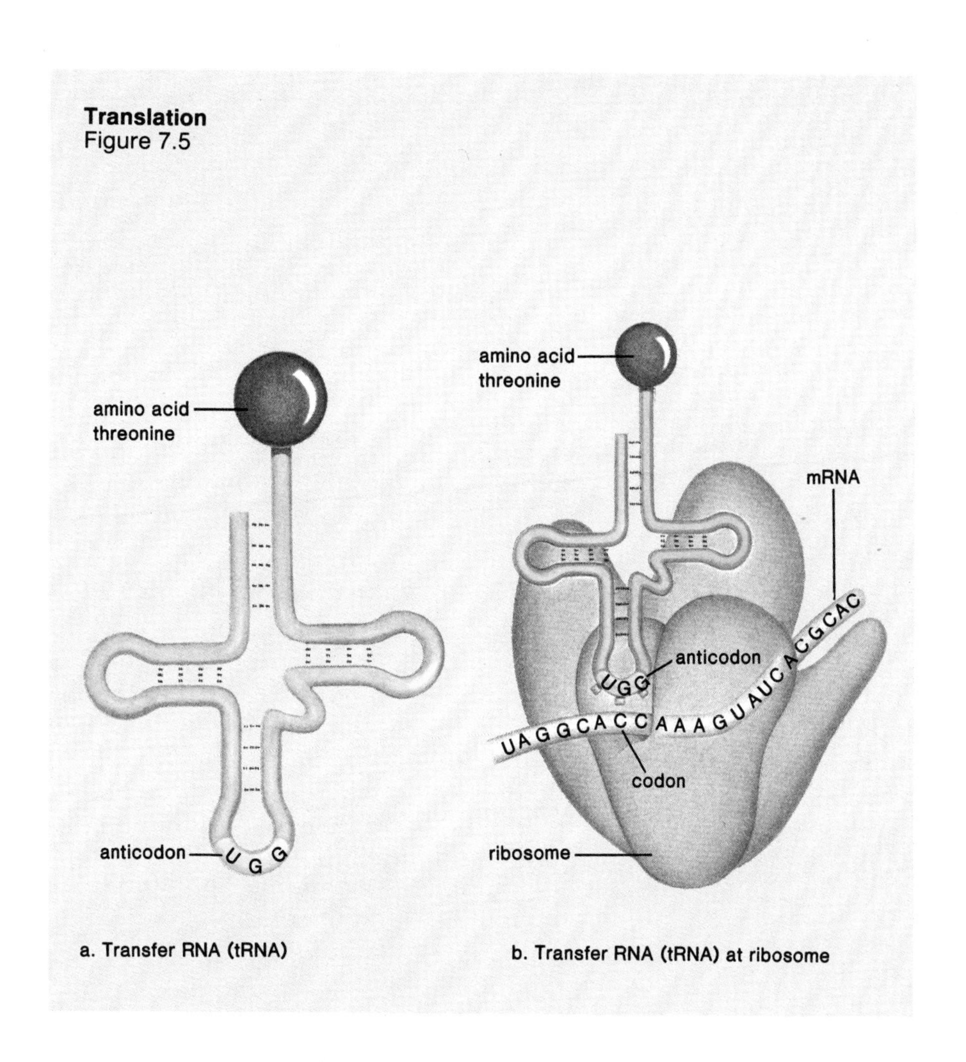

Translation
Figure 7.5
amino acid
threonine
anticodon
U G G
amino acid
threonine
mRNA
anticodon
UGG
UAGGCACCAAAGUAUCACGCAC
codon
ribosome

a. Transfer RNA (tRNA)

b. Transfer RNA (tRNA) at ribosome

Name of Molecule	Special Significance	Definition	Name of Molecule	Special Significance	Definition
DNA	Code	Sequence of 3 DNA bases	rRNA	Ribosome	Site of protein synthesis
mRNA	Codon	Sequence of 3 RNA bases complementary to DNA	Amino acid	Building block for protein	Transported to ribosome by tRNA
tRNA	Anticodon	Sequence of 3 RNA bases complementary to mRNA codon	Protein	Enzymes and certain cell components	Amino acids joined in a predetermined order

a.

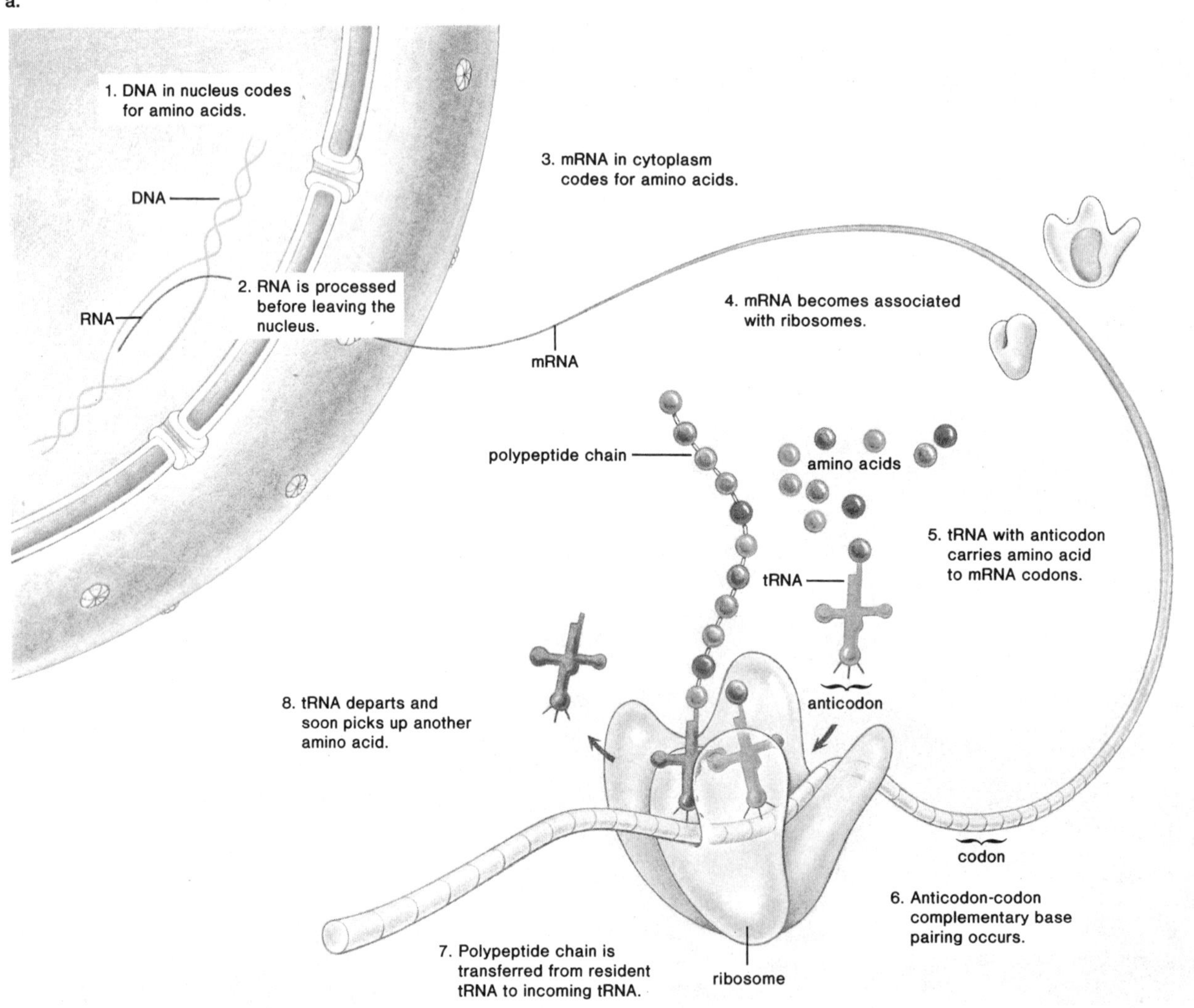
1. DNA in nucleus codes for amino acids.
DNA
2. RNA is processed before leaving the nucleus.
RNA
3. mRNA in cytoplasm codes for amino acids.
4. mRNA becomes associated with ribosomes.
mRNA
polypeptide chain
amino acids
tRNA
5. tRNA with anticodon carries amino acid to mRNA codons.
anticodon
8. tRNA departs and soon picks up another amino acid.
codon
6. Anticodon-codon complementary base pairing occurs.
7. Polypeptide chain is transferred from resident tRNA to incoming tRNA.
ribosome

b.

Summary of Protein Synthesis
Figure 7.6

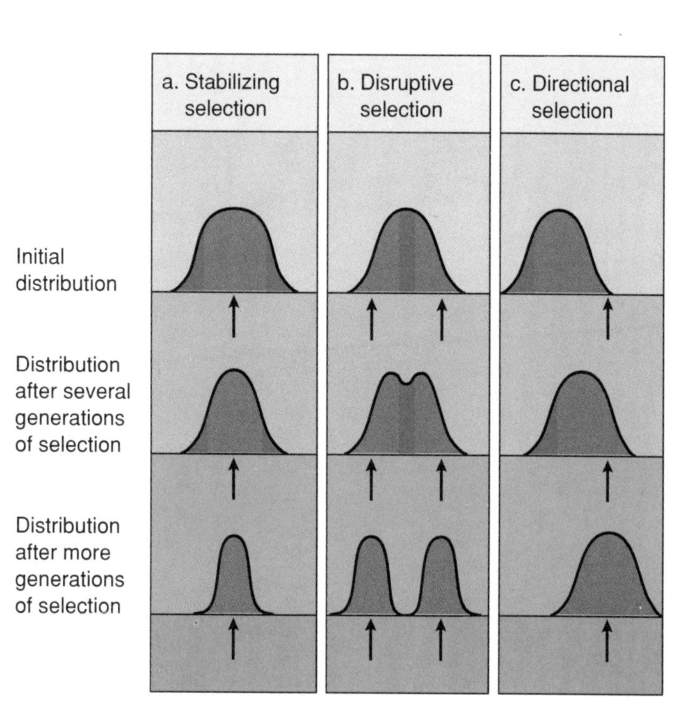
a. Stabilizing selection
b. Disruptive selection
c. Directional selection
Initial distribution
Distribution after several generations of selection
Distribution after more generations of selection

Types of Natural Selection
Figure 8.9

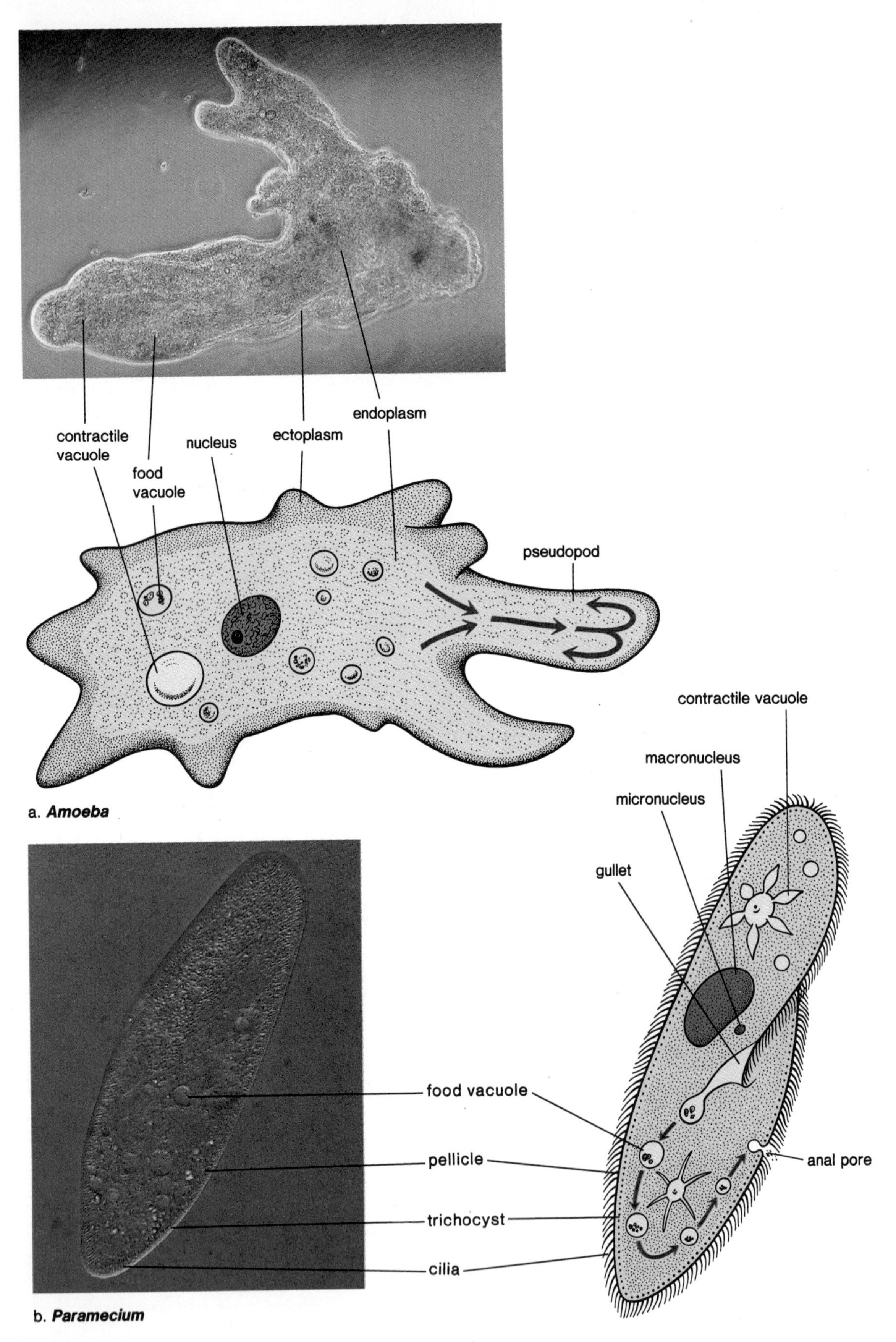
contractile
vacuole
food
vacuole
nucleus
ectoplasm
endoplasm
pseudopod
a. Amoeba
contractile vacuole
macronucleus
micronucleus
gullet
food vacuole
pellicle
trichocyst
cilia
anal pore
b. Paramecium

***Amoeba* vs. *Paramecium* Anatomy**
Figure 10.5

***Chlamydomonas* Life Cycle**
Figure 10.7

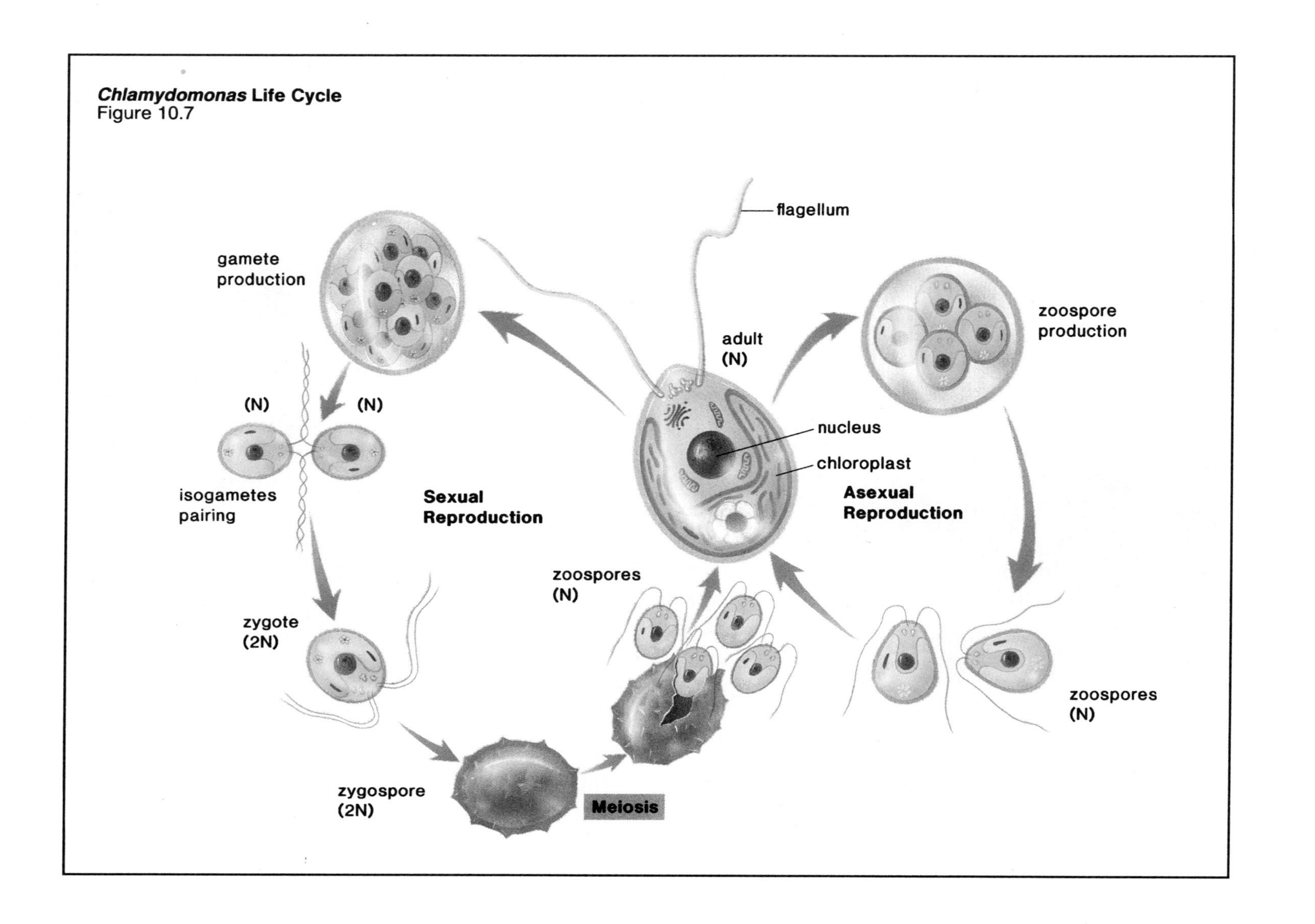
flagellum
gamete
production
adult
(N)
zoospore
production
(N)
(N)
nucleus
chloroplast
isogametes
pairing
Sexual
Reproduction
Asexual
Reproduction
zoospores
(N)
zygote
(2N)
zoospores
(N)
zygospore
(2N)
Meiosis

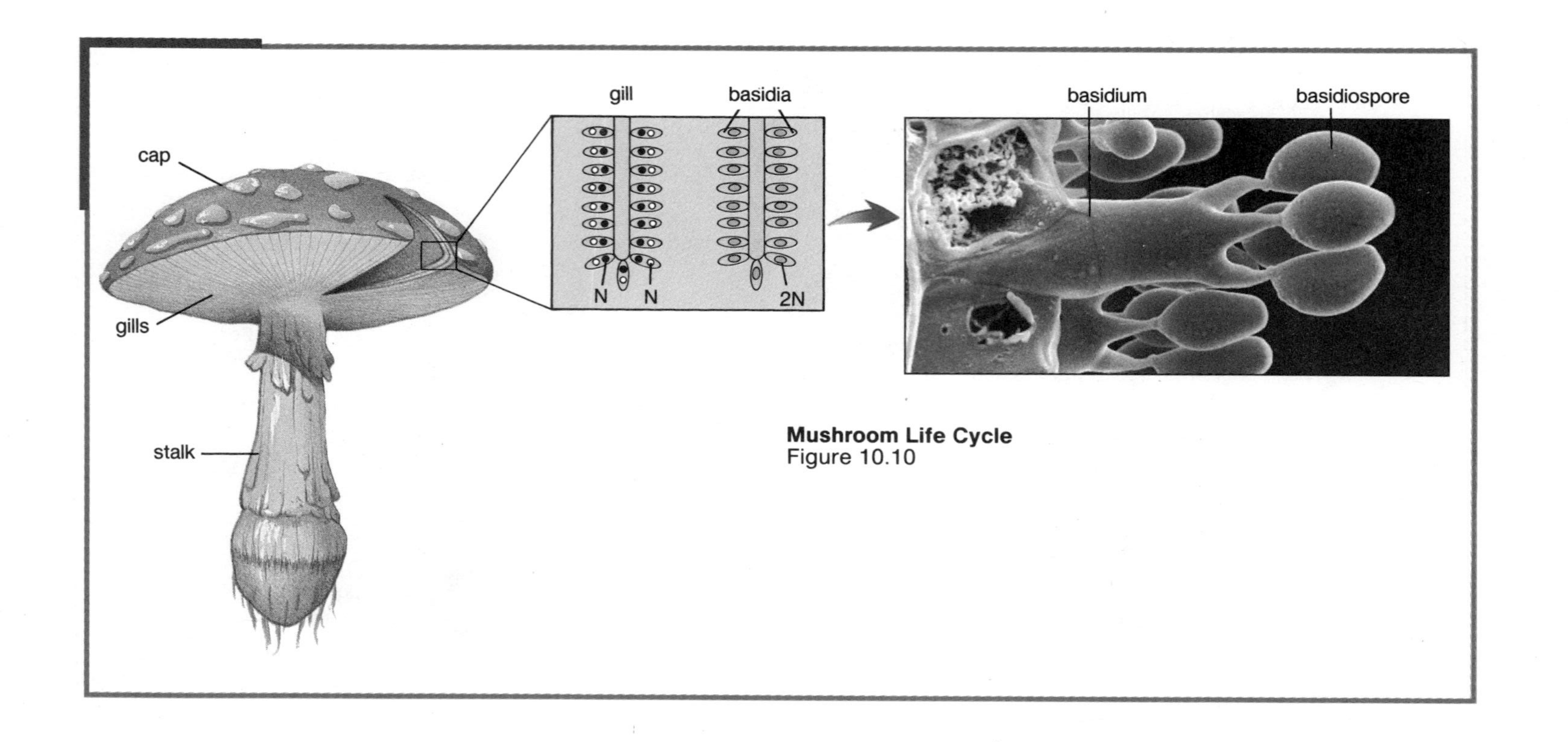
cap
gills
stalk
gill
basidia
N
N
2N
basidium
basidiospore

Mushroom Life Cycle
Figure 10.10

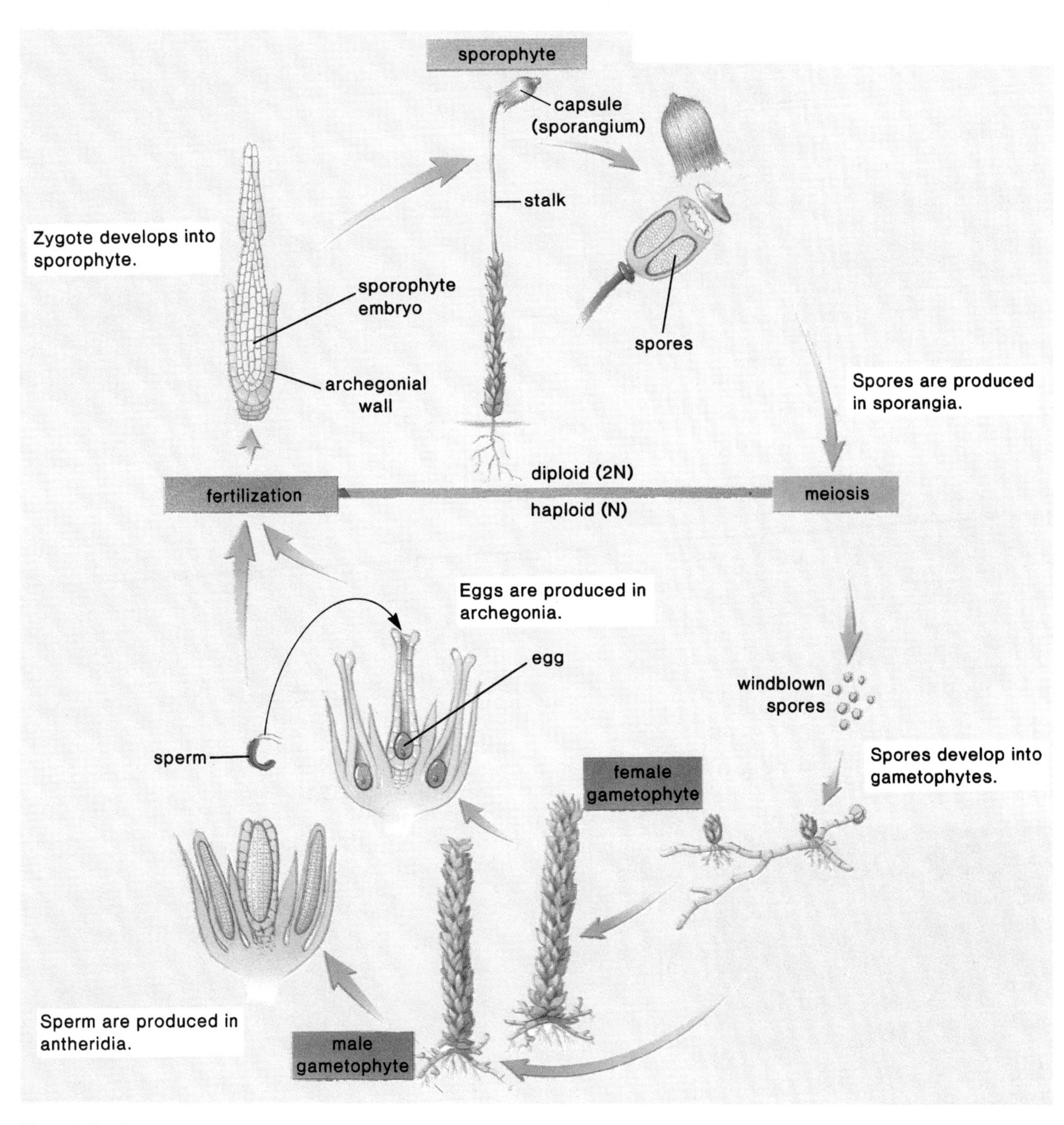
sporophyte
capsule
(sporangium)
stalk
Zygote develops into
sporophyte.
sporophyte
embryo
spores
archegonial
wall
Spores are produced
in sporangia.
diploid (2N)
fertilization
meiosis
haploid (N)
Eggs are produced in
archegonia.
egg
windblown
spores
sperm
Spores develop into
gametophytes.
female
gametophyte
Sperm are produced in
antheridia.
male
gametophyte

Moss Life Cycle
Figure 11.3b

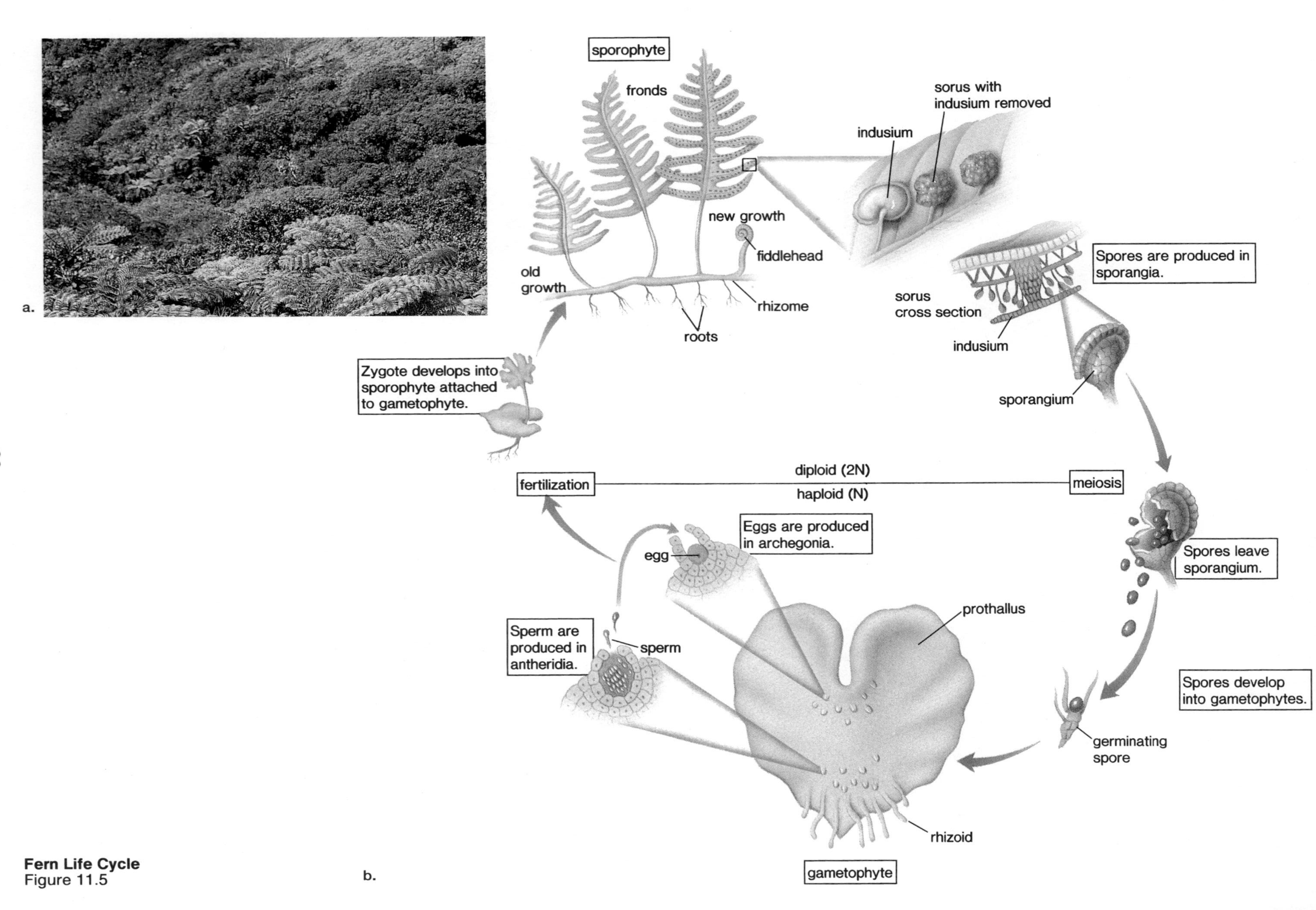
a.
sporophyte
fronds
sorus with
indusium removed
indusium
new growth
fiddlehead
old
growth
rhizome
roots
Spores are produced in
sporangia.
sorus
cross section
indusium
sporangium
Zygote develops into
sporophyte attached
to gametophyte.
diploid (2N)
haploid (N)
fertilization
meiosis
Eggs are produced
in archegonia.
egg
Spores leave
sporangium.
prothallus
Sperm are
produced in
antheridia.
sperm
Spores develop
into gametophytes.
germinating
spore
rhizoid
gametophyte
b.

Fern Life Cycle
Figure 11.5

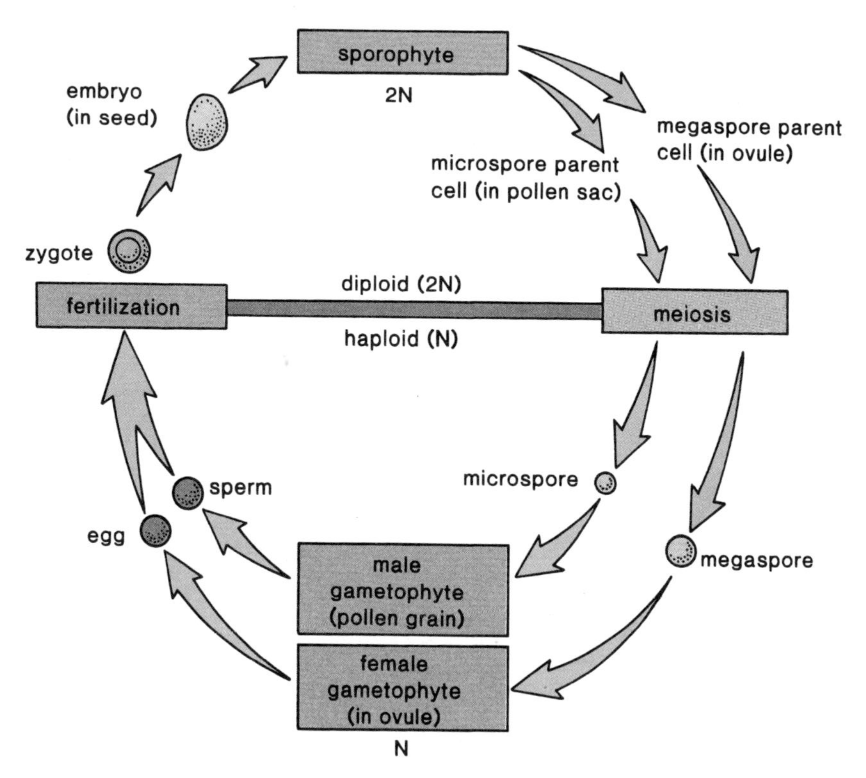
sporophyte
2N
embryo
(in seed)
megaspore parent
cell (in ovule)
microspore parent
cell (in pollen sac)
zygote
diploid (2N)
fertilization
meiosis
haploid (N)
microspore
sperm
egg
megaspore
male
gametophyte
(pollen grain)
female
gametophyte
(in ovule)
N

Seed Plant Life Cycle
Figure 11.6

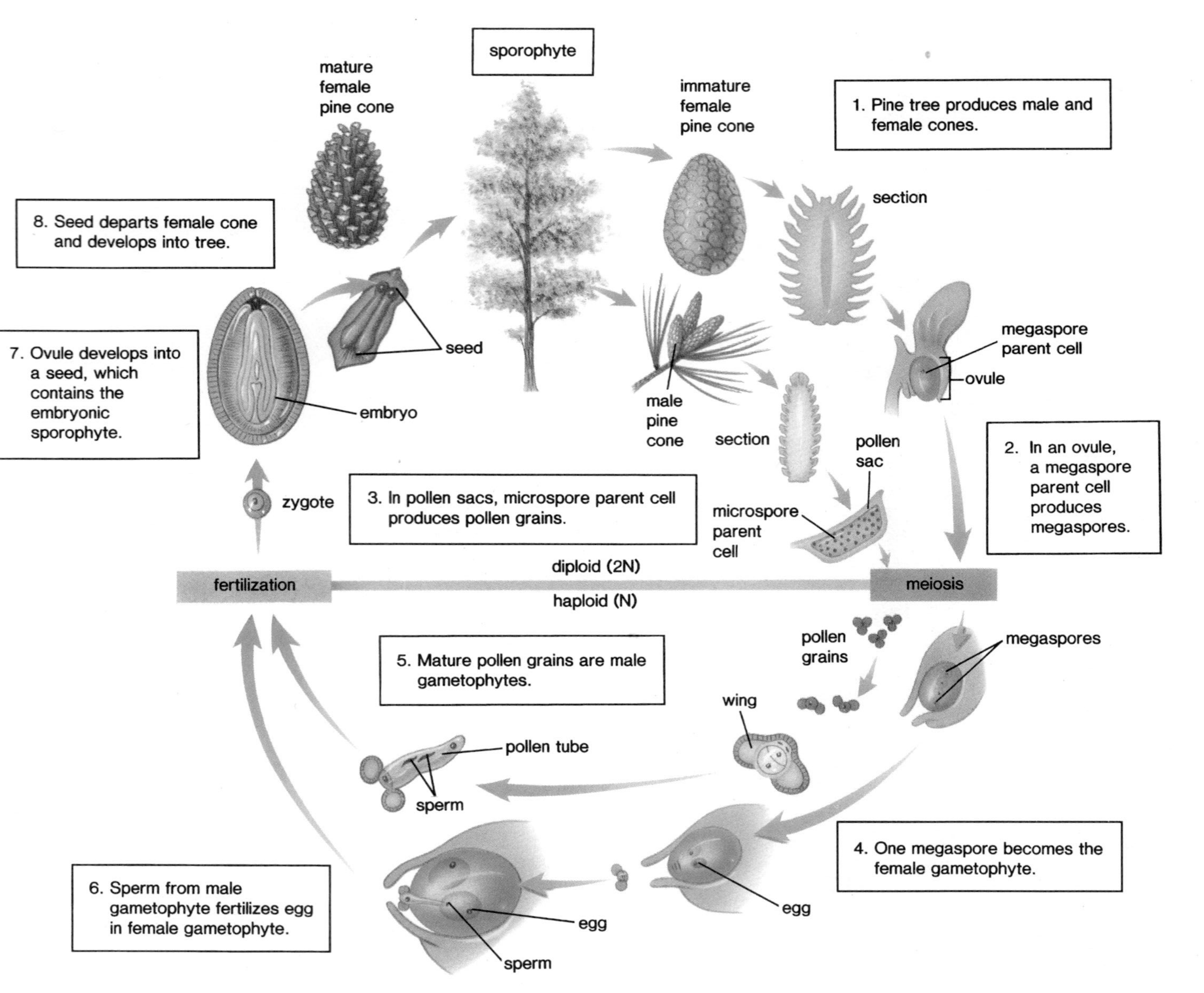

sporophyte
mature female pine cone
immature female pine cone
1. Pine tree produces male and female cones.
section
8. Seed departs female cone and develops into tree.
seed
megaspore parent cell
ovule
7. Ovule develops into a seed, which contains the embryonic sporophyte.
embryo
male pine cone
section
pollen sac
2. In an ovule, a megaspore parent cell produces megaspores.
zygote
3. In pollen sacs, microspore parent cell produces pollen grains.
microspore parent cell
diploid (2N)
fertilization
meiosis
haploid (N)
pollen grains
megaspores
5. Mature pollen grains are male gametophytes.
wing
pollen tube
sperm
4. One megaspore becomes the female gametophyte.
6. Sperm from male gametophyte fertilizes egg in female gametophyte.
egg
egg
sperm

Pine Life Cycle
Figure 11.7

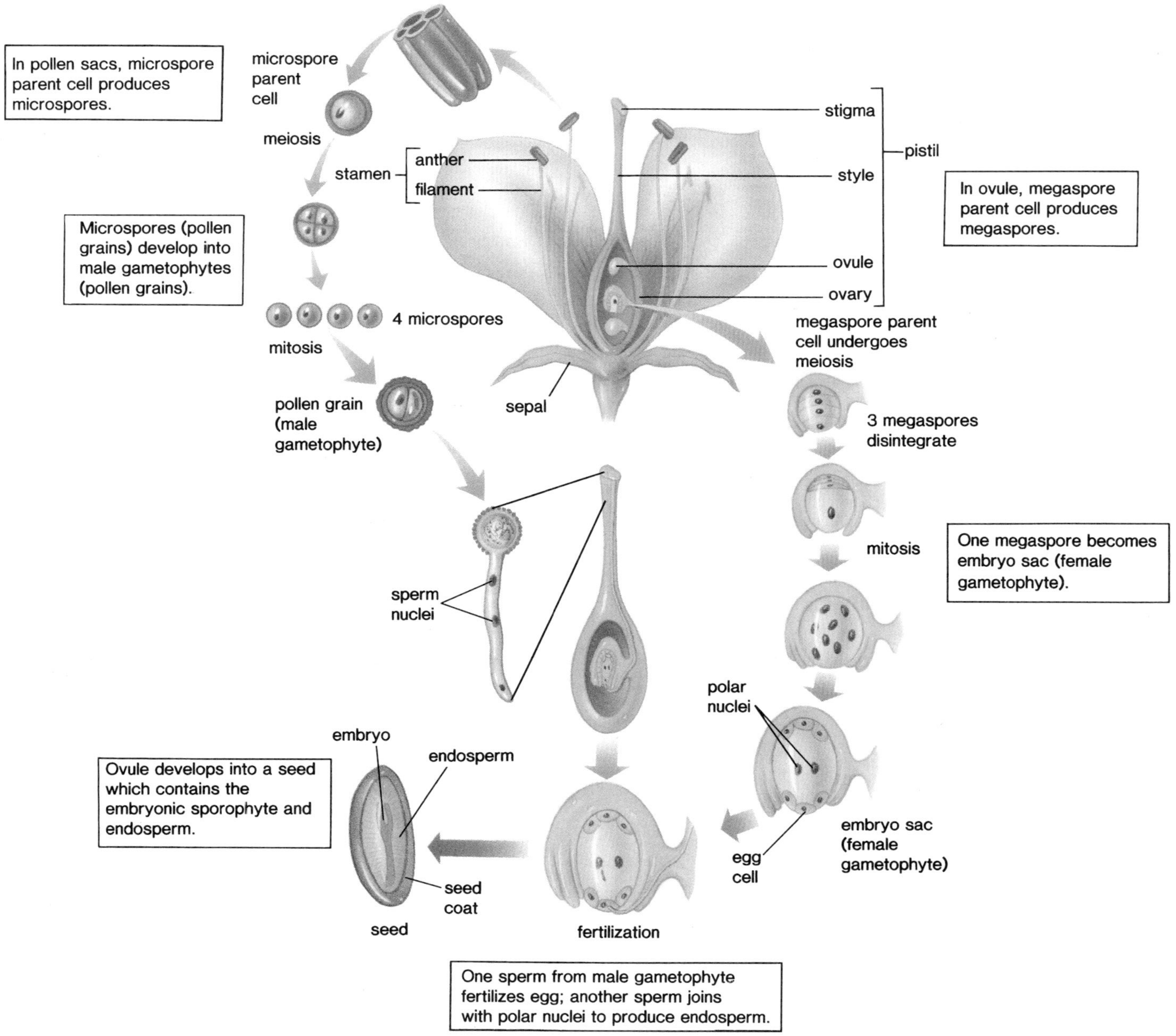

In pollen sacs, microspore parent cell produces microspores.

microspore parent cell

meiosis

stamen

anther

filament

Microspores (pollen grains) develop into male gametophytes (pollen grains).

4 microspores

mitosis

pollen grain (male gametophyte)

sepal

stigma

style

ovule

ovary

pistil

In ovule, megaspore parent cell produces megaspores.

megaspore parent cell undergoes meiosis

3 megaspores disintegrate

mitosis

One megaspore becomes embryo sac (female gametophyte).

sperm nuclei

polar nuclei

egg cell

embryo sac (female gametophyte)

embryo

endosperm

seed coat

seed

Ovule develops into a seed which contains the embryonic sporophyte and endosperm.

fertilization

One sperm from male gametophyte fertilizes egg; another sperm joins with polar nuclei to produce endosperm.

Flowering Plant Life Cycle
Figure 11.9

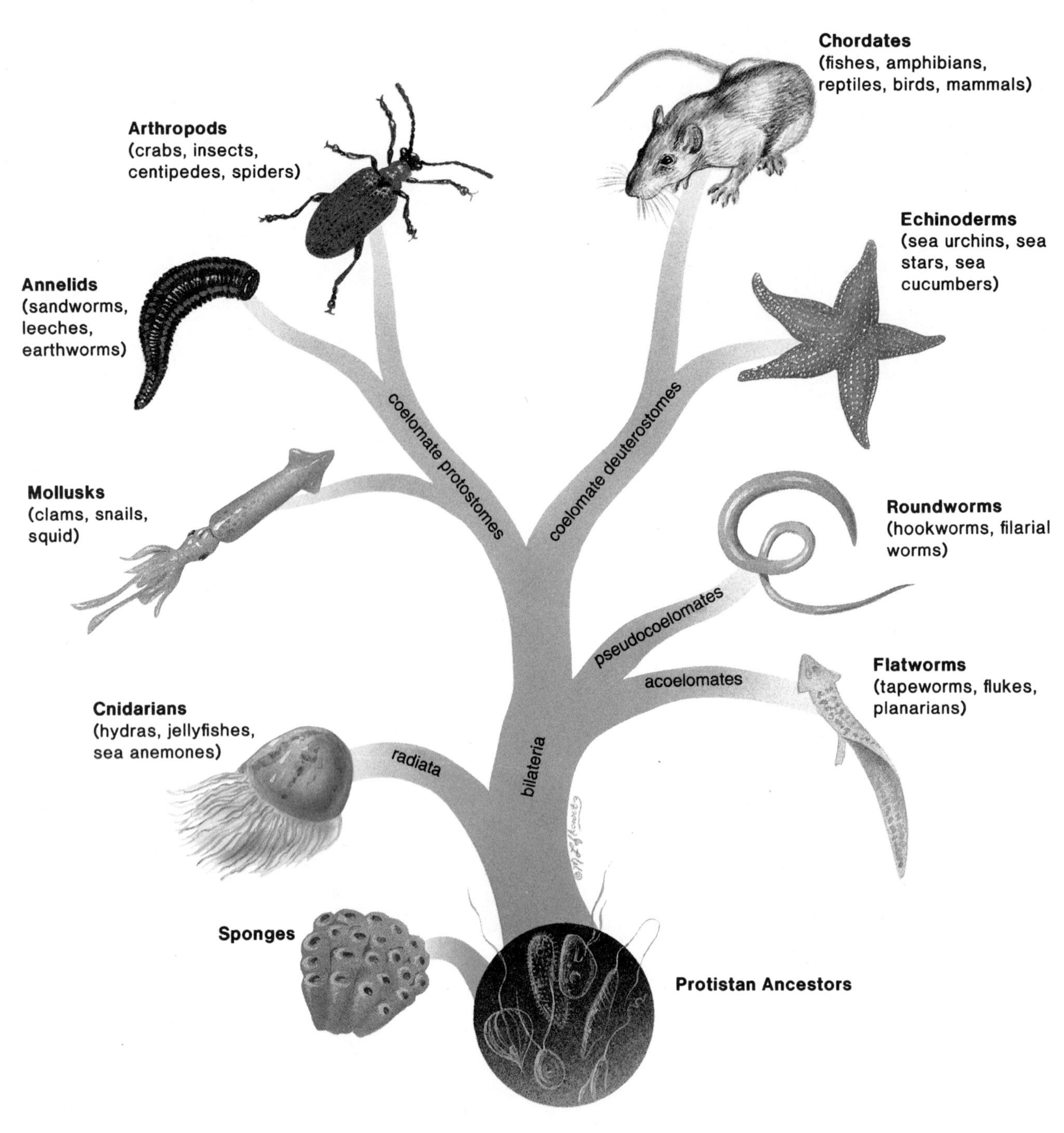
Chordates
(fishes, amphibians, reptiles, birds, mammals)
Arthropods
(crabs, insects, centipedes, spiders)
Echinoderms
(sea urchins, sea stars, sea cucumbers)
Annelids
(sandworms, leeches, earthworms)
coelomate protostomes
coelomate deuterostomes
Mollusks
(clams, snails, squid)
Roundworms
(hookworms, filarial worms)
pseudocoelomates
acoelomates
Flatworms
(tapeworms, flukes, planarians)
Cnidarians
(hydras, jellyfishes, sea anemones)
radiata
bilateria
Sponges
Protistan Ancestors

Evolutionary Tree of the Animal Kingdom
Figure 12.2

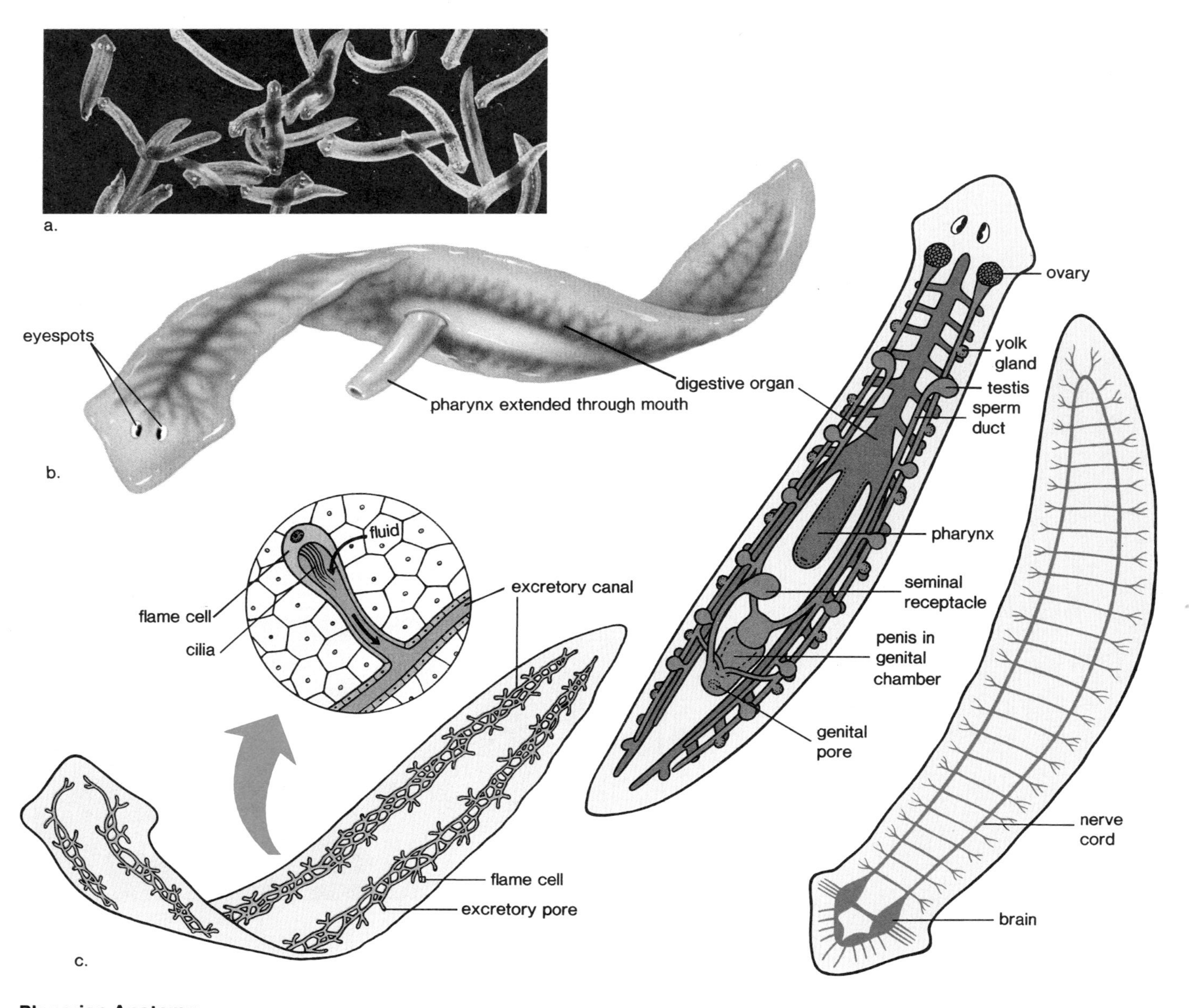

a.
eyespots
digestive organ
pharynx extended through mouth
b.
ovary
yolk gland
testis
sperm duct
pharynx
seminal receptacle
penis in genital chamber
genital pore
nerve cord
brain
fluid
flame cell
cilia
excretory canal
flame cell
excretory pore
c.

Planarian Anatomy
Figure 12.5

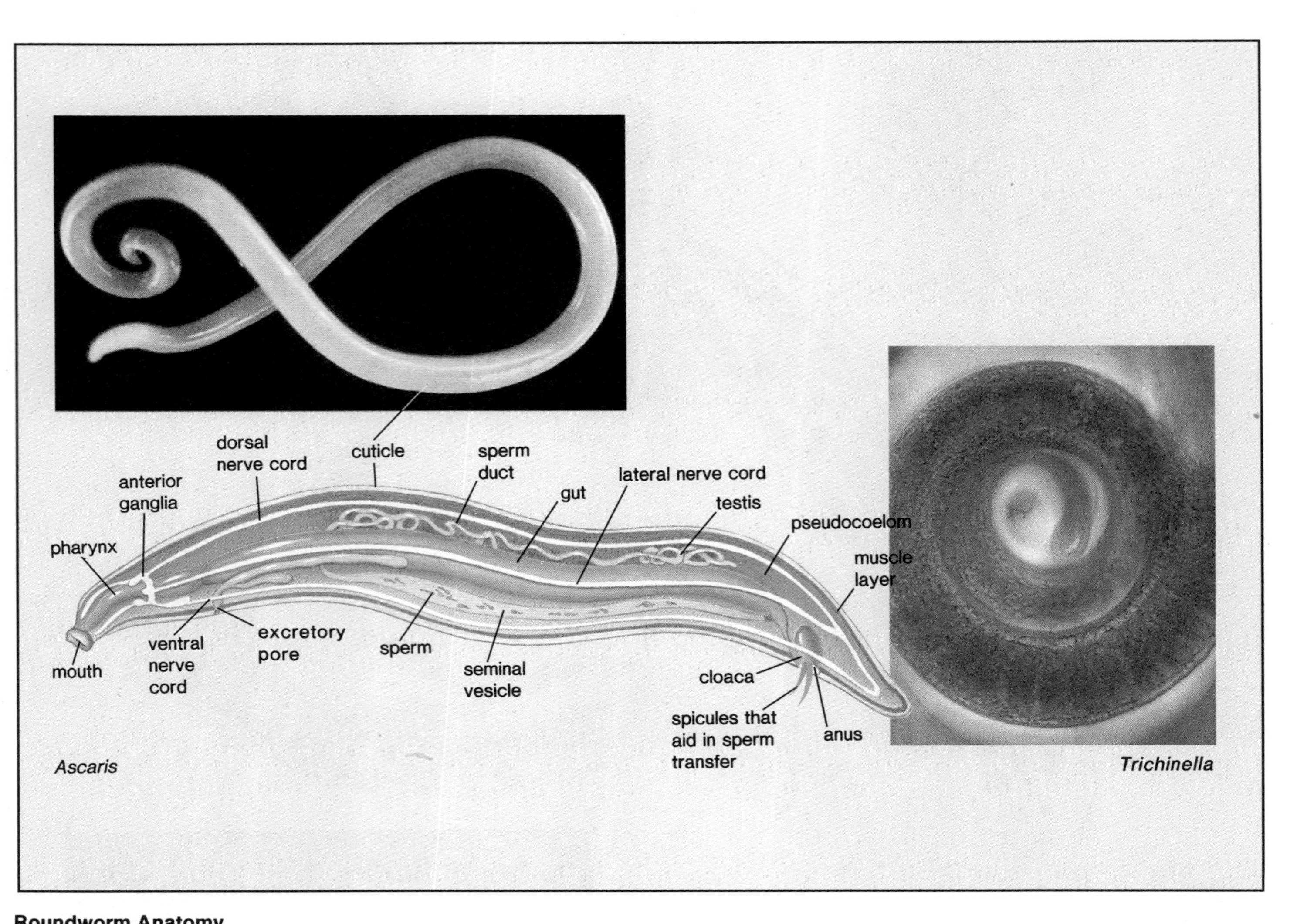
dorsal nerve cord
cuticle
sperm duct
lateral nerve cord
gut
testis
anterior ganglia
pharynx
pseudocoelom
muscle layer
mouth
ventral nerve cord
excretory pore
sperm
seminal vesicle
cloaca
spicules that aid in sperm transfer
anus
Ascaris
Trichinella

Roundworm Anatomy
Figure 12.6

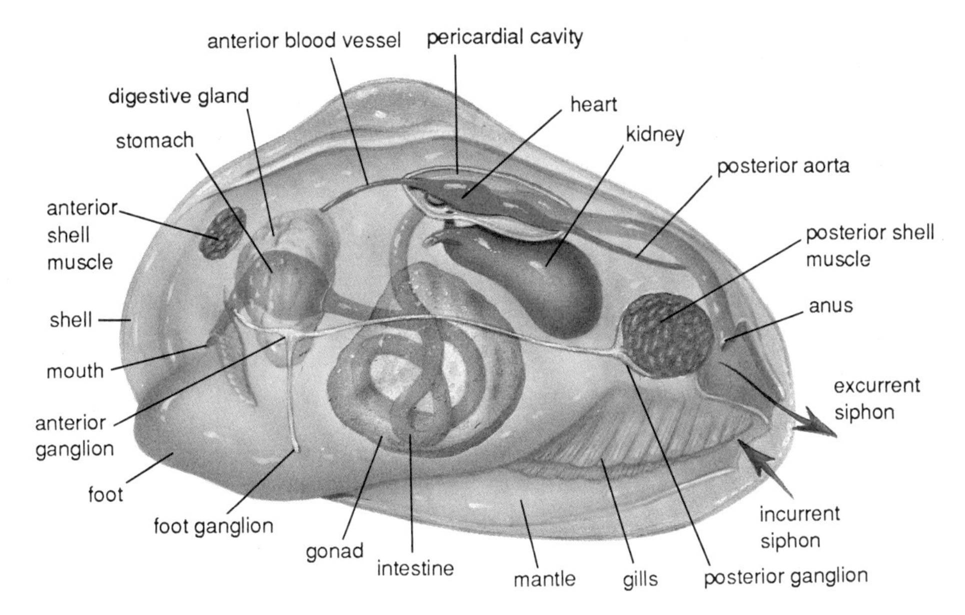
anterior blood vessel
pericardial cavity
digestive gland
heart
stomach
kidney
posterior aorta
anterior shell muscle
posterior shell muscle
shell
anus
mouth
excurrent siphon
anterior ganglion
foot
incurrent siphon
foot ganglion
gonad
intestine
mantle
gills
posterior ganglion
b.

Clam Anatomy
Figure 12.7b

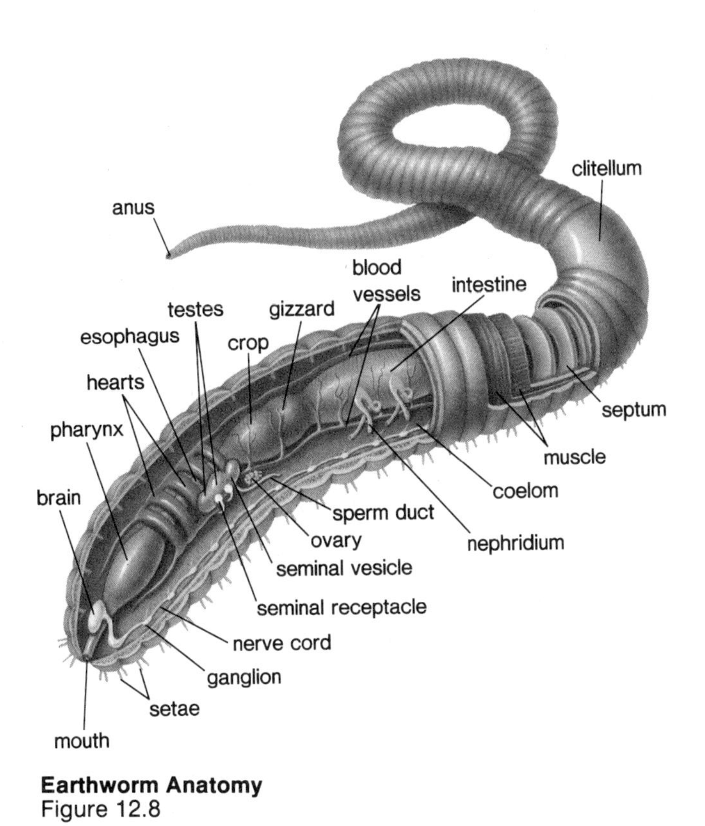
clitellum
anus
blood vessels
intestine
testes
gizzard
esophagus
crop
hearts
septum
pharynx
muscle
coelom
brain
sperm duct
ovary
nephridium
seminal vesicle
seminal receptacle
nerve cord
ganglion
setae
mouth

Earthworm Anatomy
Figure 12.8

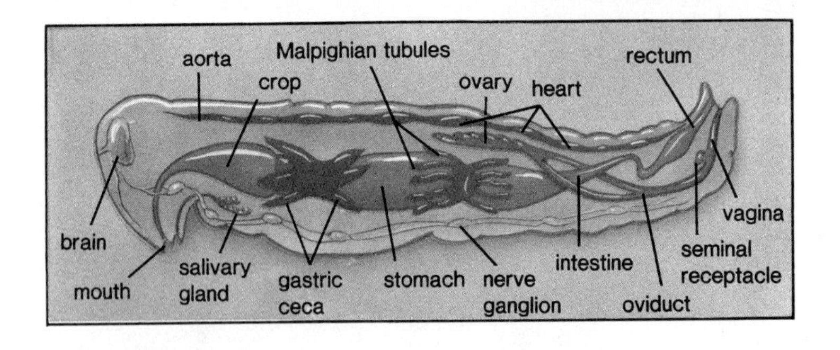
aorta
Malpighian tubules
rectum
crop
ovary
heart
vagina
brain
seminal
receptacle
salivary
gland
gastric
ceca
stomach
nerve
ganglion
intestine
oviduct
mouth

Female Grasshopper Anatomy
Figure 12.9b

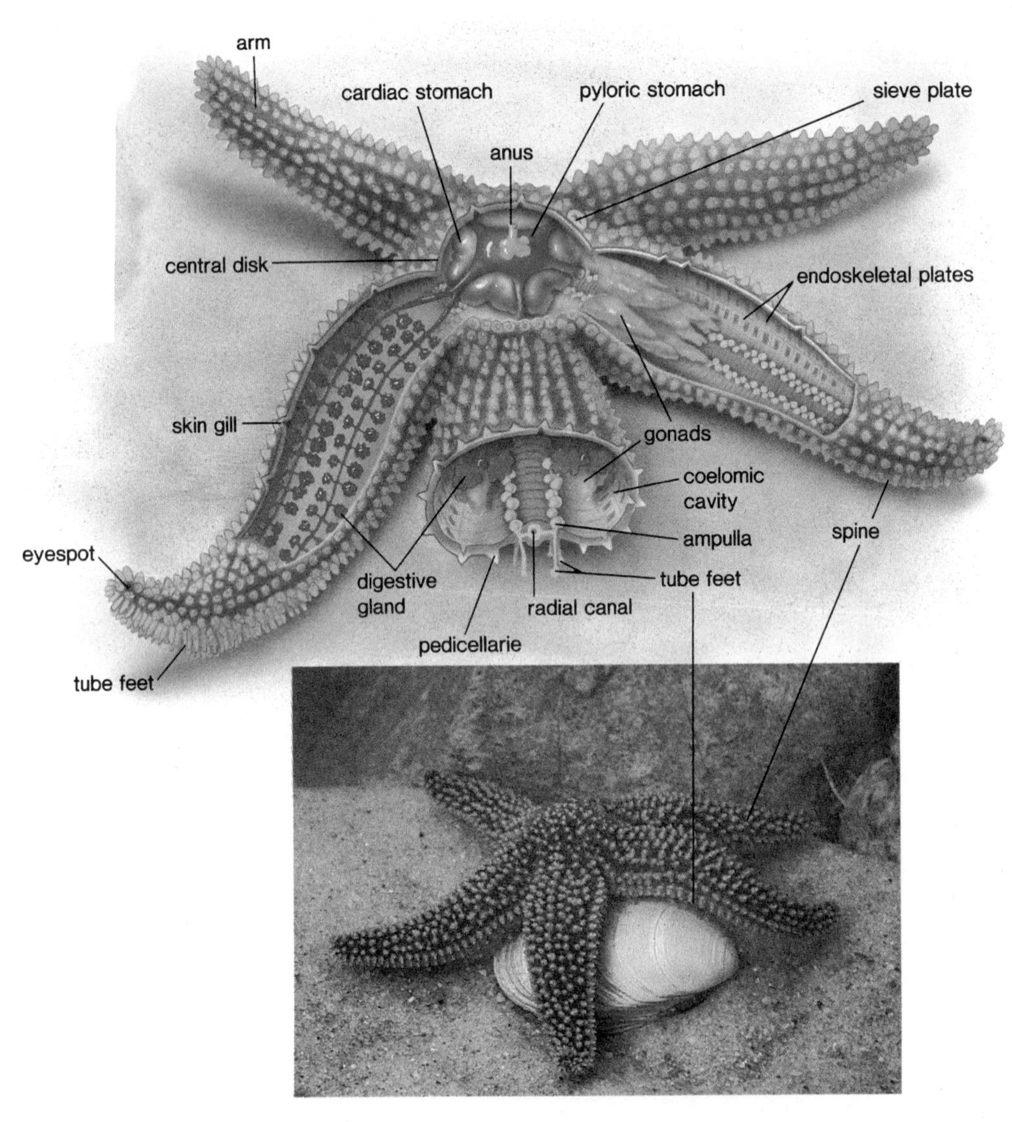
arm
cardiac stomach
pyloric stomach
sieve plate
anus
central disk
endoskeletal plates
skin gill
gonads
coelomic
cavity
ampulla
spine
eyespot
digestive
gland
tube feet
radial canal
pedicellarie
tube feet

Sea Star Anatomy
Figure 12.10b

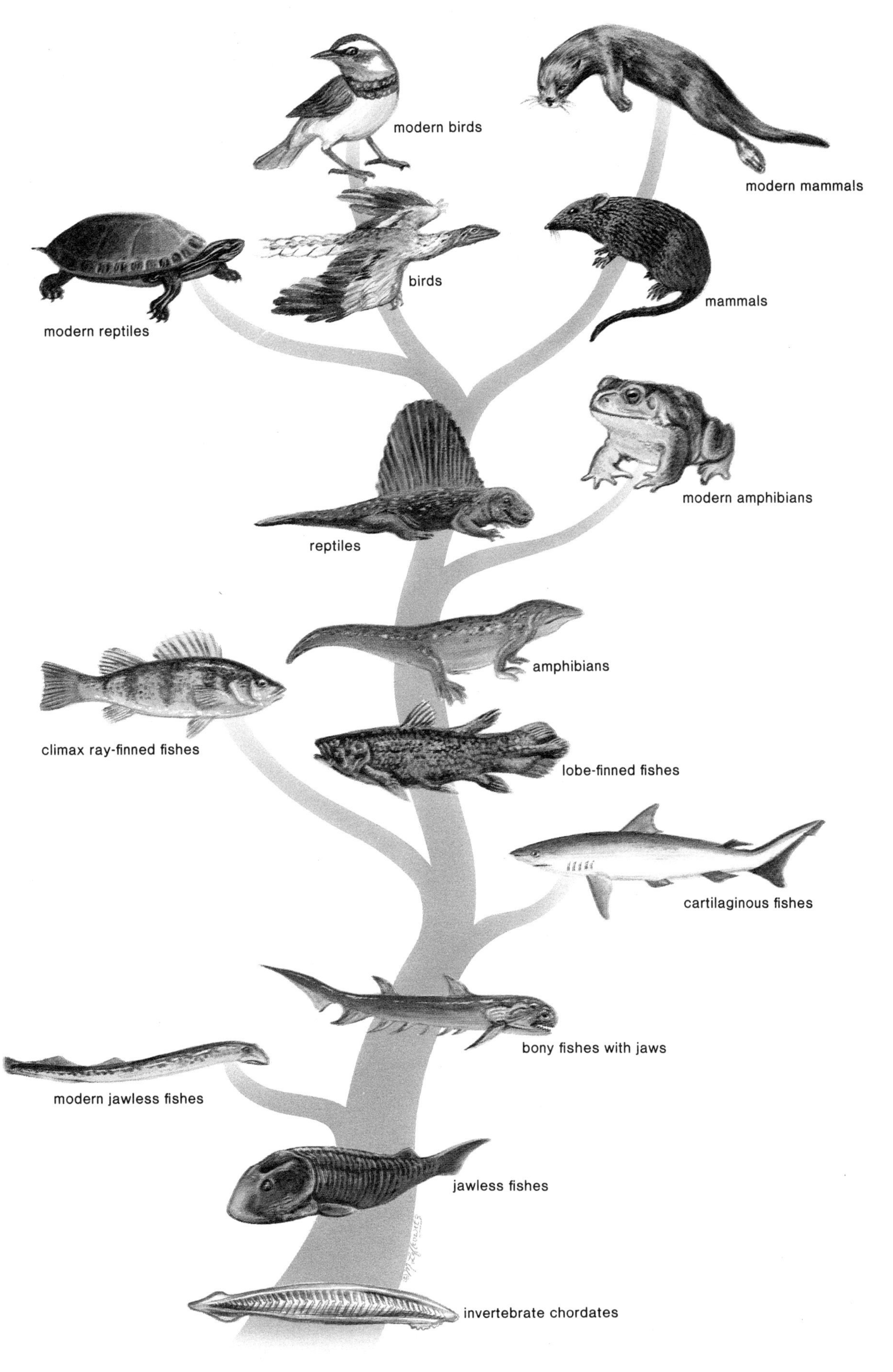
modern birds
modern mammals
birds
mammals
modern reptiles
modern amphibians
reptiles
amphibians
climax ray-finned fishes
lobe-finned fishes
cartilaginous fishes
bony fishes with jaws
modern jawless fishes
jawless fishes
invertebrate chordates

Evolutionary Tree of Vertebrates
Figure 12.12

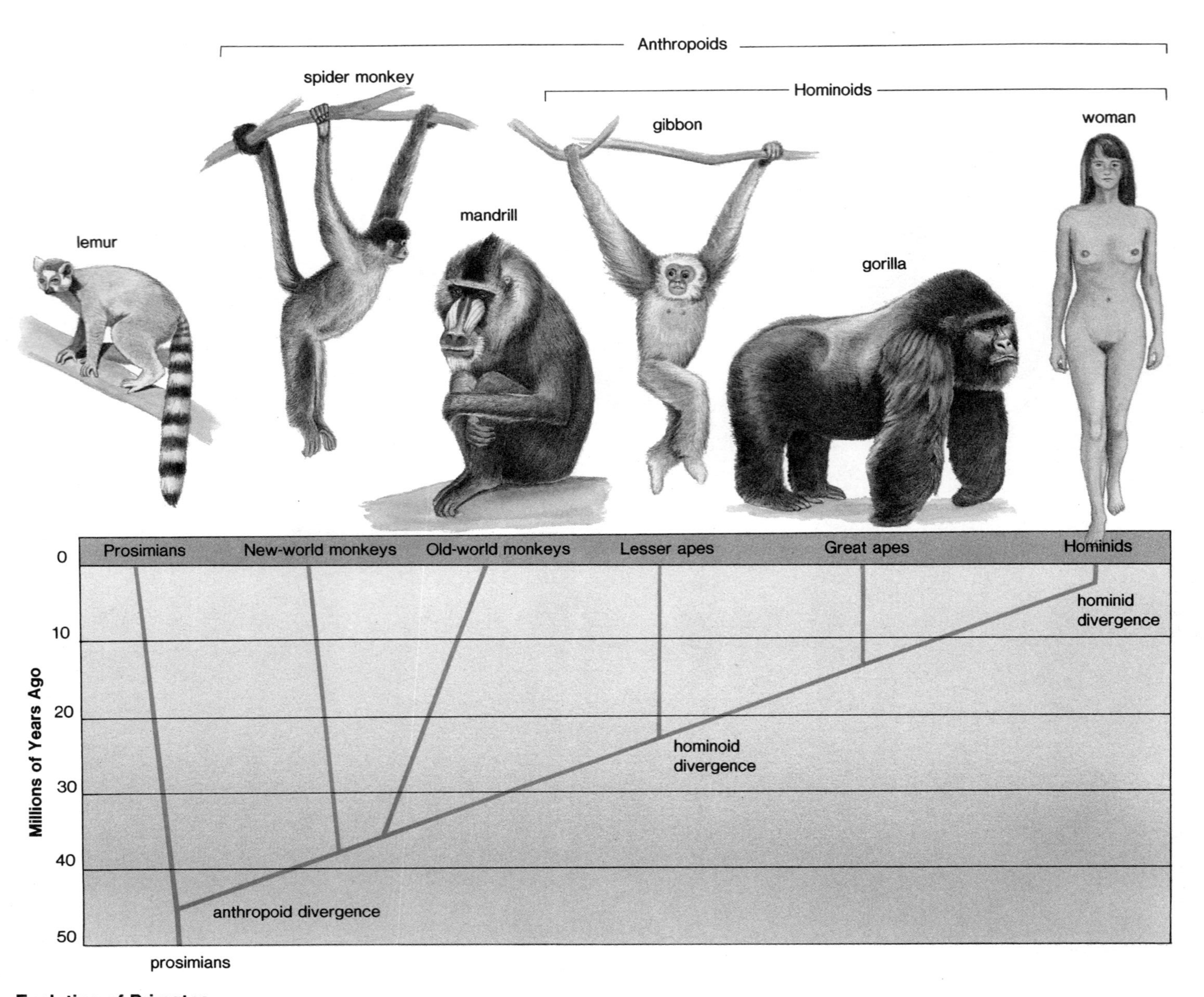
Anthropoids
Hominoids
lemur
spider monkey
mandrill
gibbon
gorilla
woman
Prosimians
New-world monkeys
Old-world monkeys
Lesser apes
Great apes
Hominids
0
10
20
30
40
50
Millions of Years Ago
hominid divergence
hominoid divergence
anthropoid divergence
prosimians

Evolution of Primates
Figure 12.15

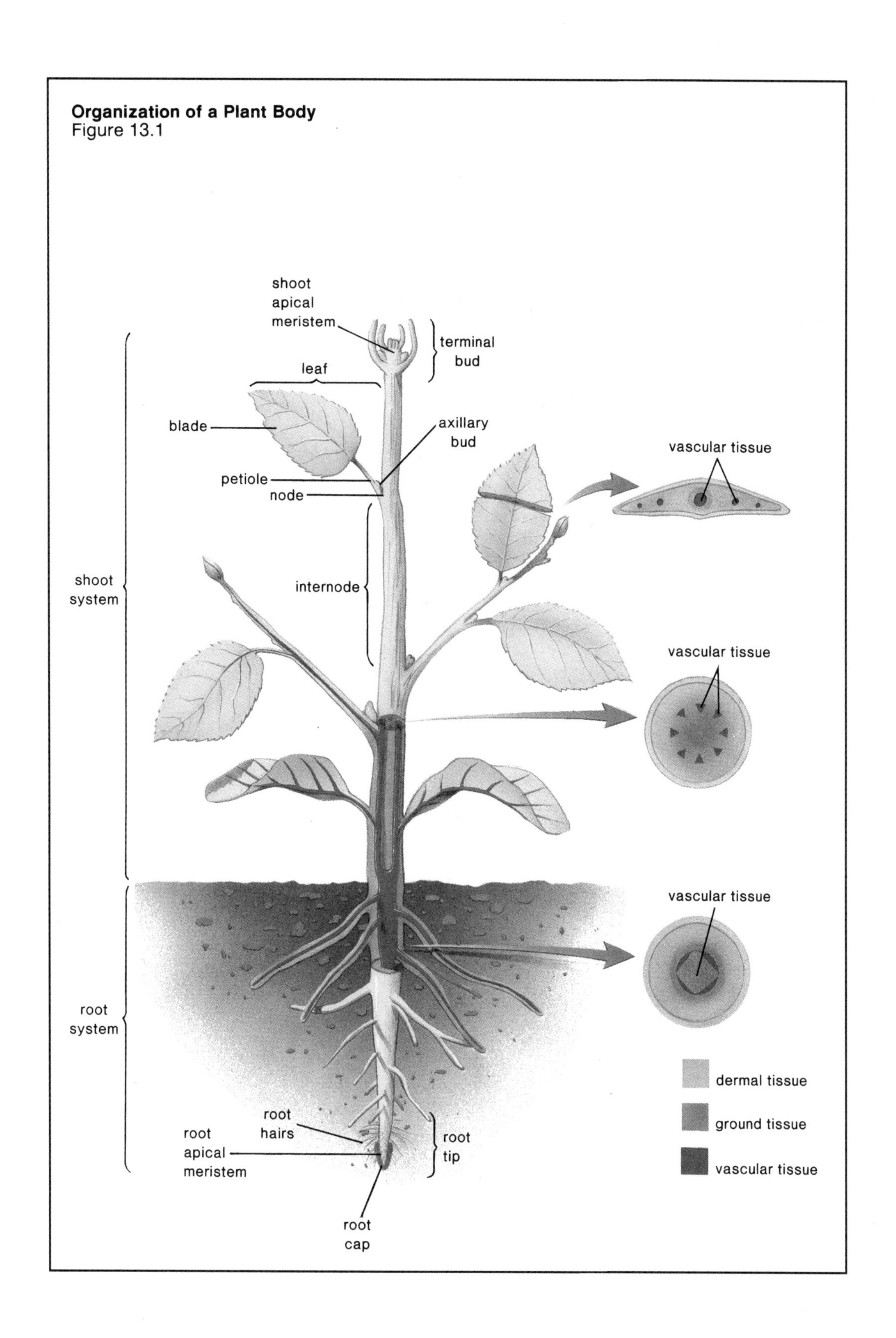
Organization of a Plant Body
Figure 13.1
shoot
apical
meristem
terminal
bud
leaf
blade
axillary
bud
petiole
node
internode
shoot
system
vascular tissue
vascular tissue
vascular tissue
root
system
root
hairs
root
apical
meristem
root
tip
root
cap
dermal tissue
ground tissue
vascular tissue

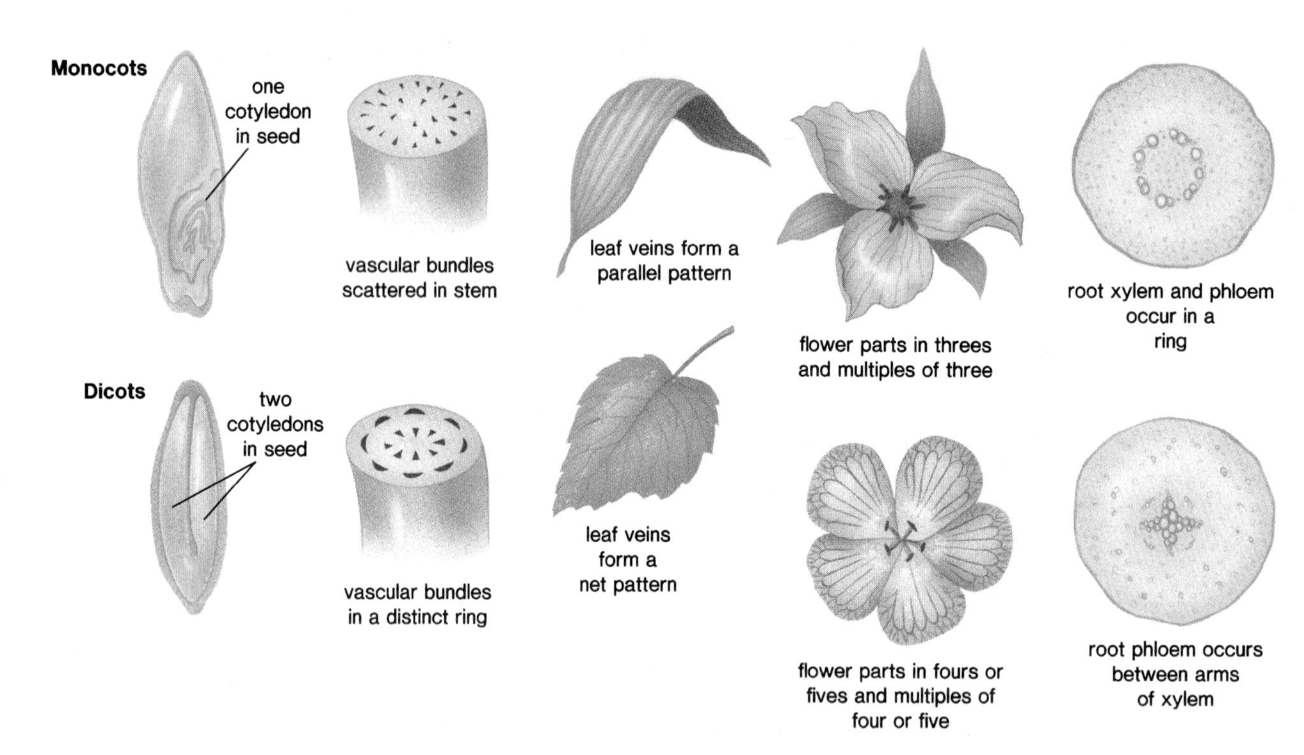
Monocots
one
cotyledon
in seed
vascular bundles
scattered in stem
leaf veins form a
parallel pattern
flower parts in threes
and multiples of three
root xylem and phloem
occur in a
ring
Dicots
two
cotyledons
in seed
vascular bundles
in a distinct ring
leaf veins
form a
net pattern
flower parts in fours or
fives and multiples of
four or five
root phloem occurs
between arms
of xylem

Monocot vs. Dicots
Figure 13.3

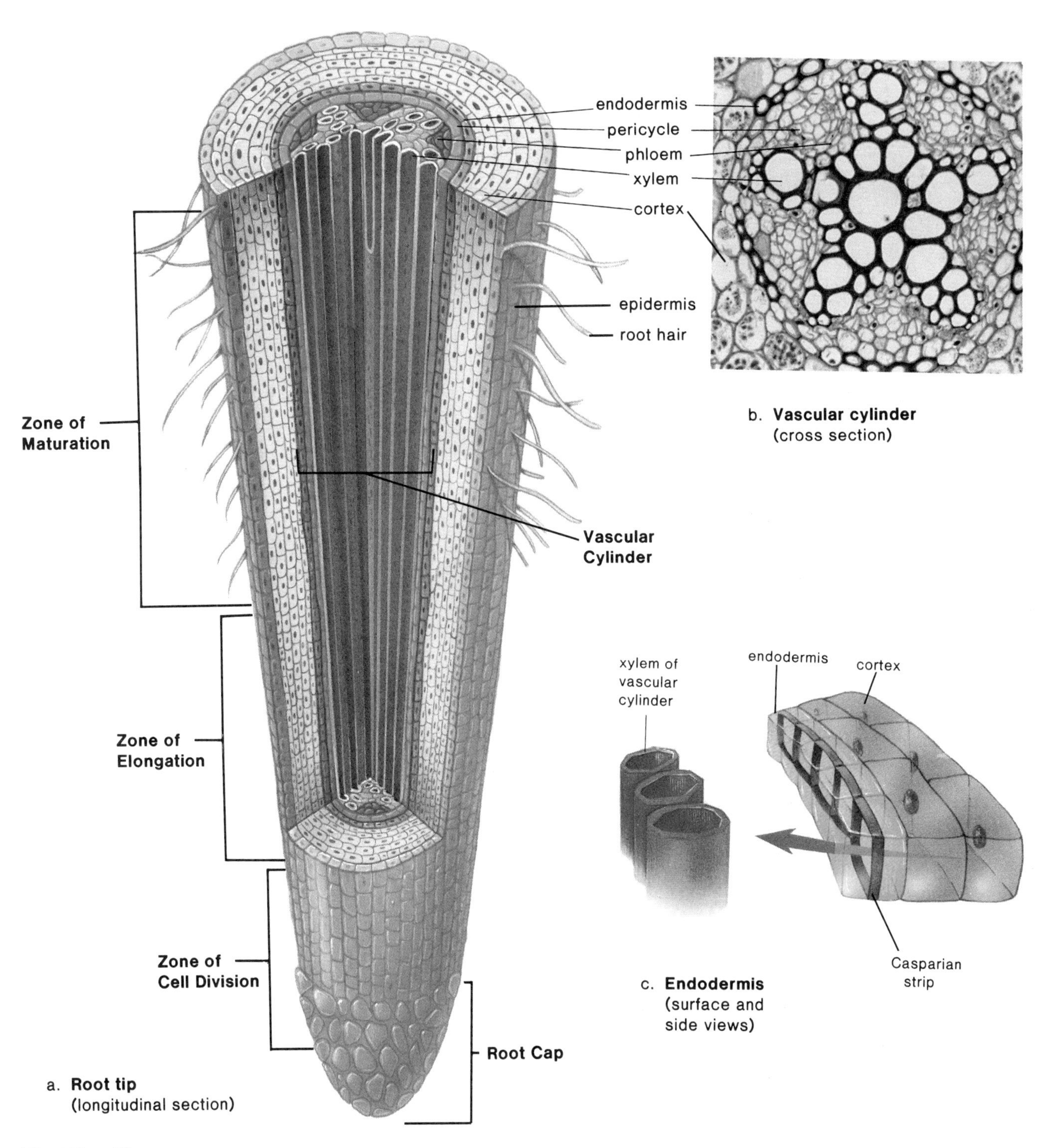
endodermis
pericycle
phloem
xylem
cortex
epidermis
root hair
Zone of Maturation
Vascular Cylinder
Zone of Elongation
Zone of Cell Division
Root Cap
xylem of vascular cylinder
endodermis
cortex
Casparian strip

a. **Root tip** (longitudinal section)

b. **Vascular cylinder** (cross section)

c. **Endodermis** (surface and side views)

Dicot Root Tip
Figure 13.4

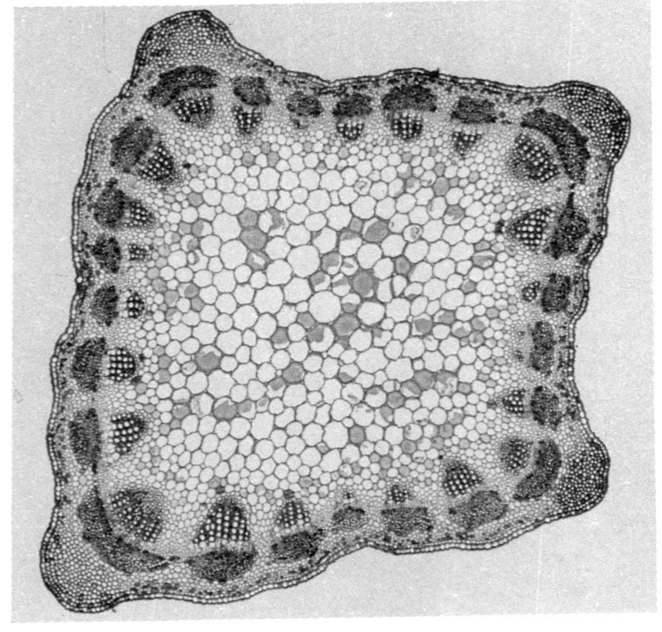
a.

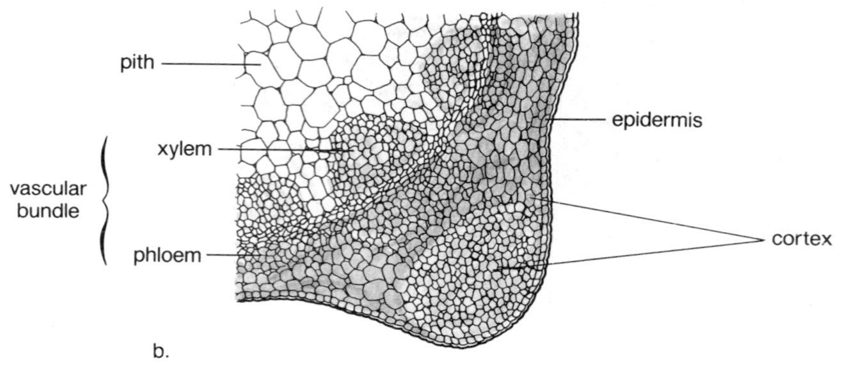
pith
xylem
vascular bundle
phloem
epidermis
cortex
b.

Herbaceous Dicot Stem Anatomy
Figure 13.6

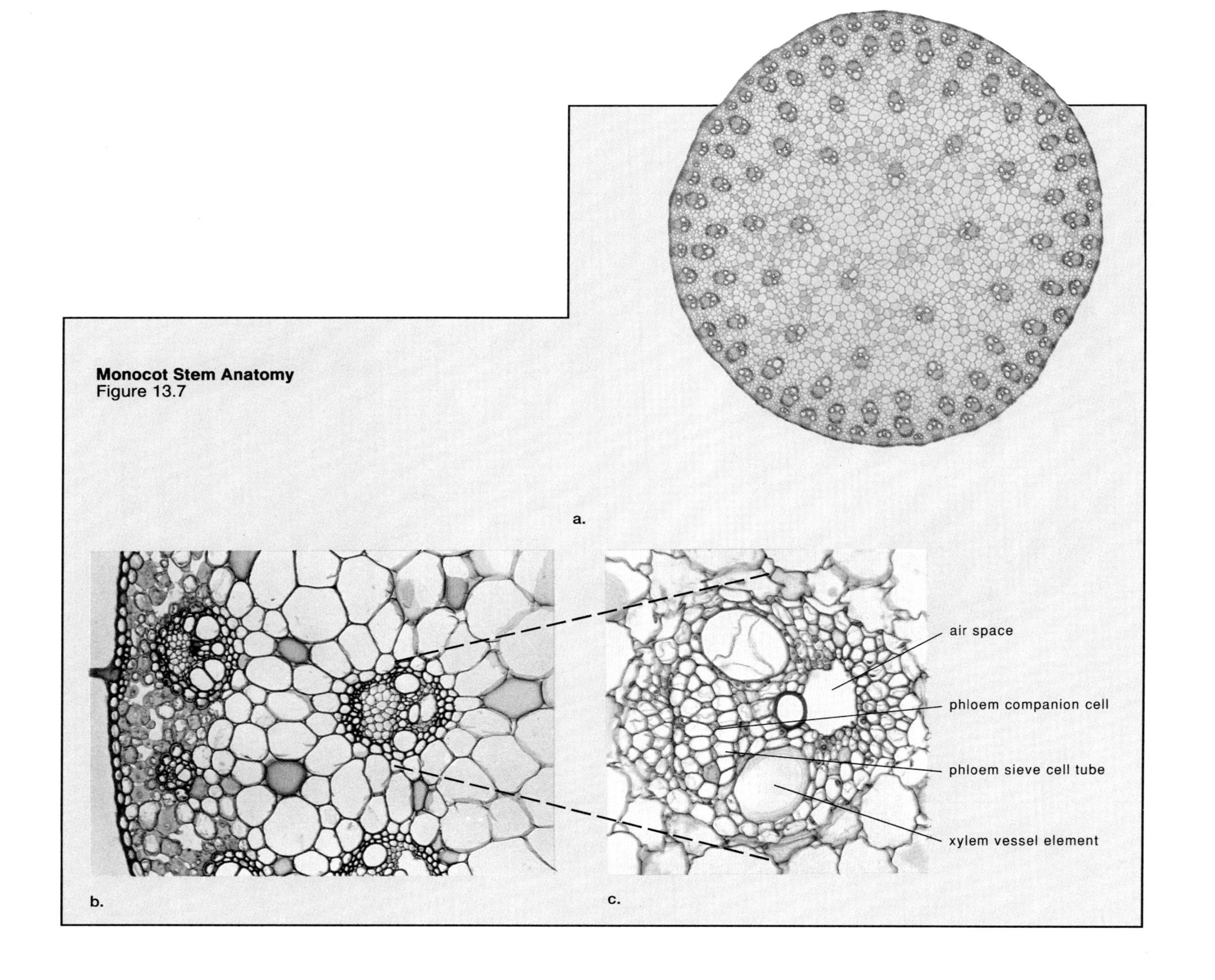
a.
air space
phloem companion cell
phloem sieve cell tube
xylem vessel element
b.
c.

Monocot Stem Anatomy
Figure 13.7

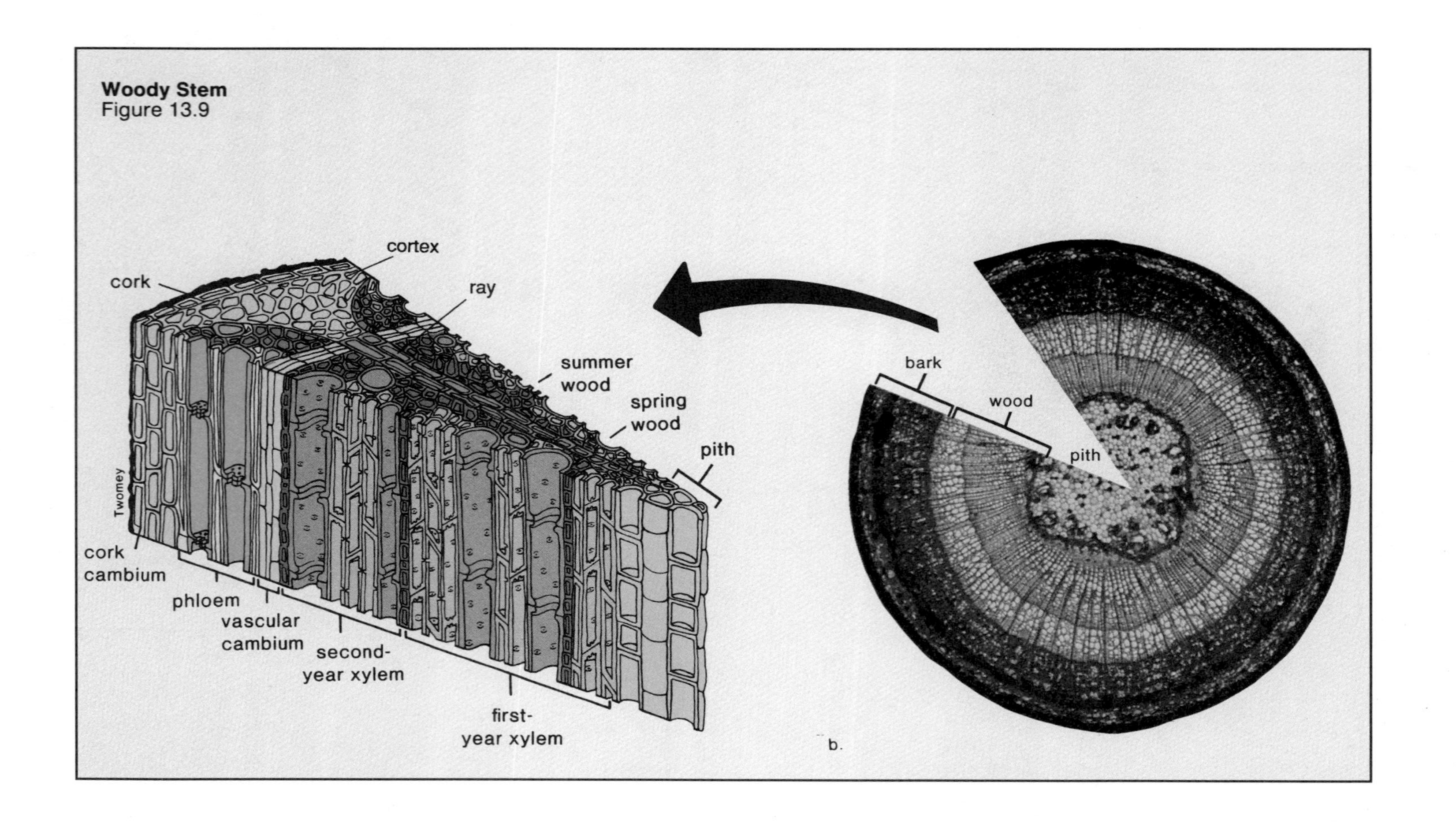
cortex
cork
ray
summer wood
spring wood
pith
Twomey
cork cambium
phloem
vascular cambium
second-year xylem
first-year xylem
bark
wood
pith
b.

Woody Stem
Figure 13.9

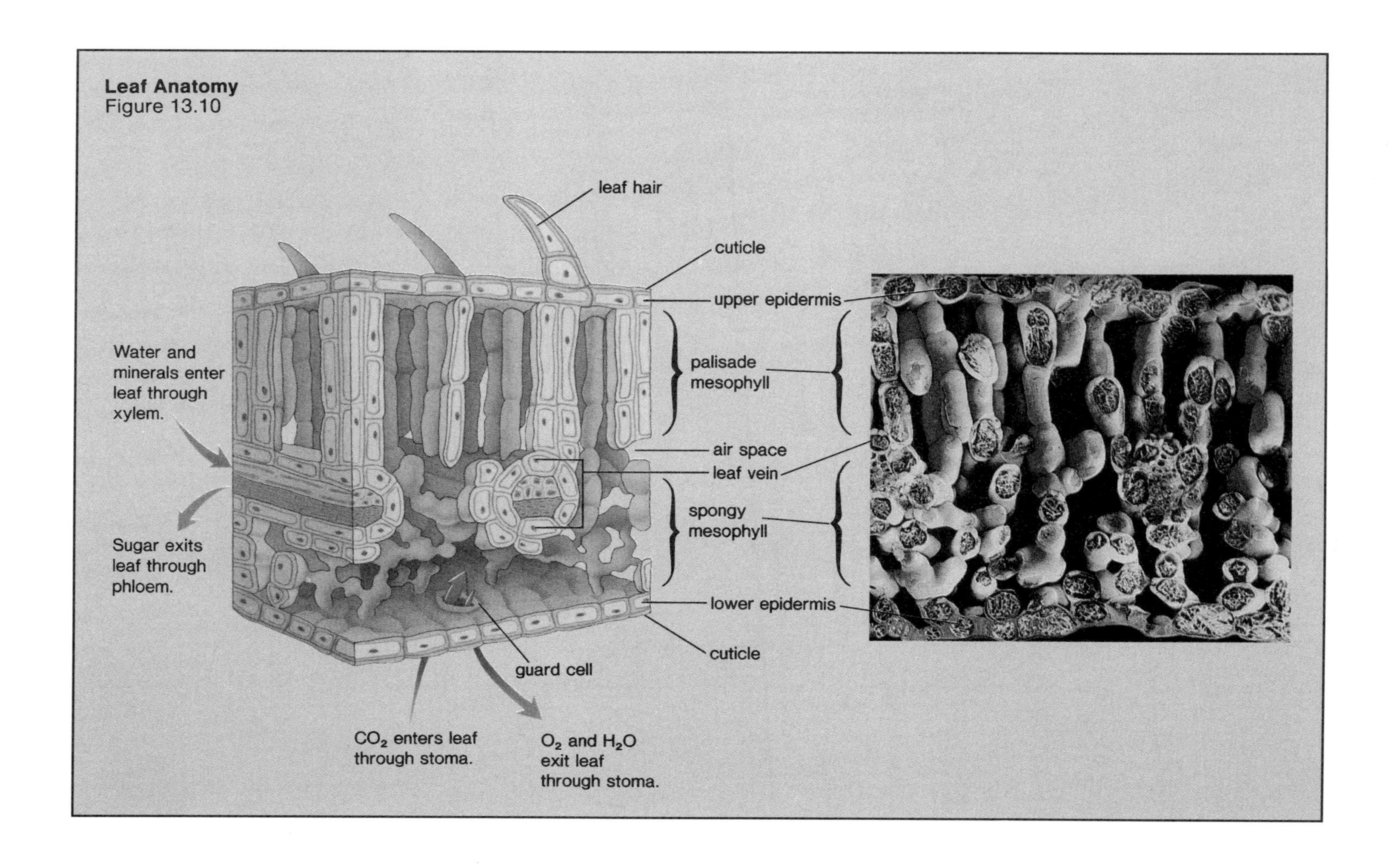
Leaf Anatomy
Figure 13.10
leaf hair
cuticle
upper epidermis
palisade mesophyll
air space
leaf vein
spongy mesophyll
lower epidermis
cuticle
guard cell
Water and minerals enter leaf through xylem.
Sugar exits leaf through phloem.
CO_2 enters leaf through stoma.
O_2 and H_2O exit leaf through stoma.

Opening and Closing of Stomata
Figure 14.3

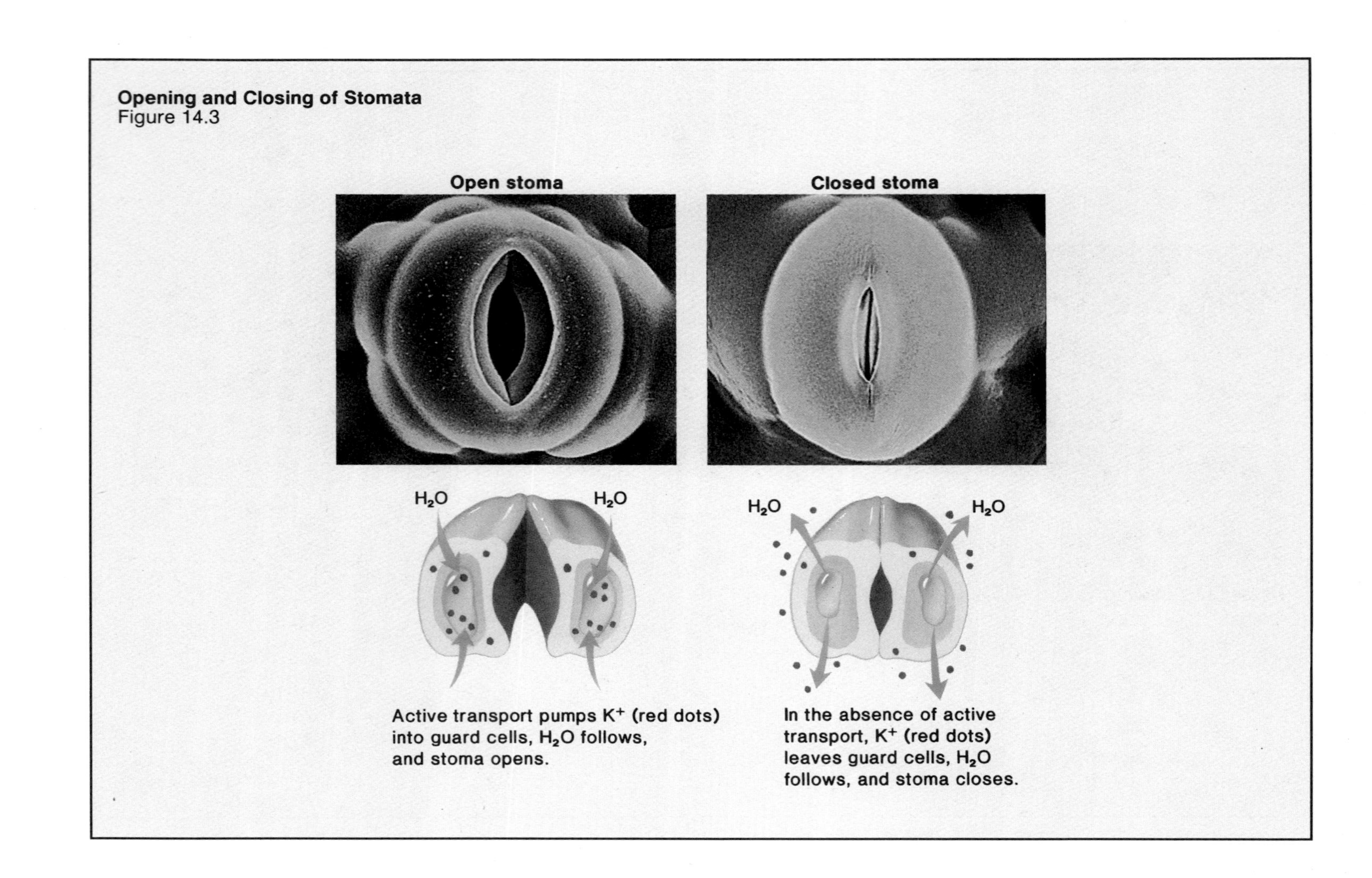
Open stoma
Closed stoma
H_2O
H_2O
H_2O
H_2O
Active transport pumps K^+ (red dots) into guard cells, H_2O follows, and stoma opens.
In the absence of active transport, K^+ (red dots) leaves guard cells, H_2O follows, and stoma closes.

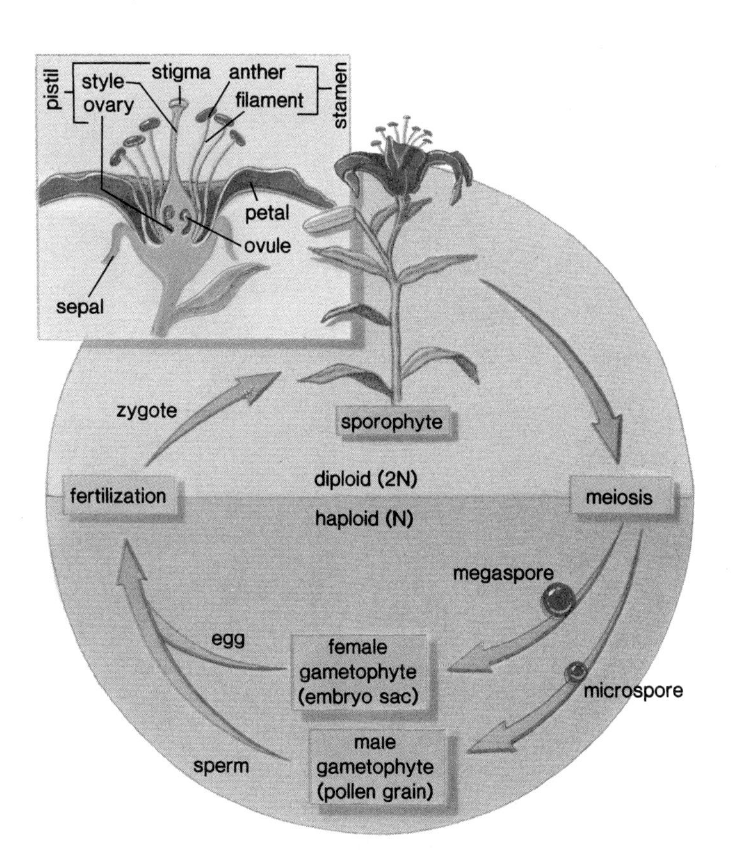
pistil
style
ovary
stigma
anther
filament
stamen
petal
ovule
sepal
zygote
sporophyte
diploid (2N)
fertilization
meiosis
haploid (N)
megaspore
egg
female
gametophyte
(embryo sac)
microspore
male
gametophyte
(pollen grain)
sperm

Alternation of Generations
Figure 14.15

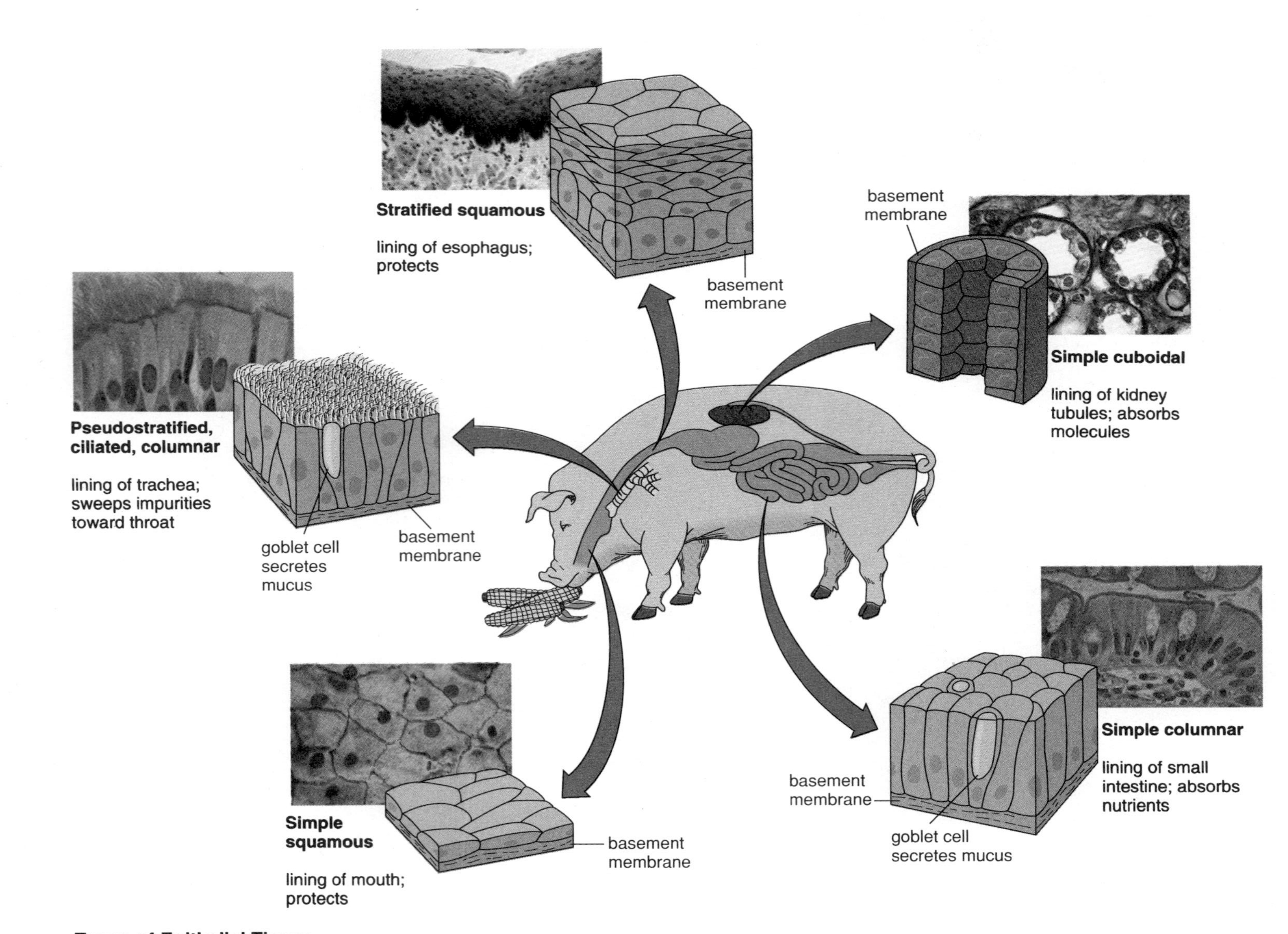
Stratified squamous
lining of esophagus; protects
basement membrane
basement membrane
Simple cuboidal
lining of kidney tubules; absorbs molecules
Pseudostratified, ciliated, columnar
lining of trachea; sweeps impurities toward throat
goblet cell secretes mucus
basement membrane
Simple squamous
lining of mouth; protects
basement membrane
basement membrane
goblet cell secretes mucus
Simple columnar
lining of small intestine; absorbs nutrients

Types of Epithelial Tissue
Figure 15.1

Anatomy of a Long Bone
Figure 15.3

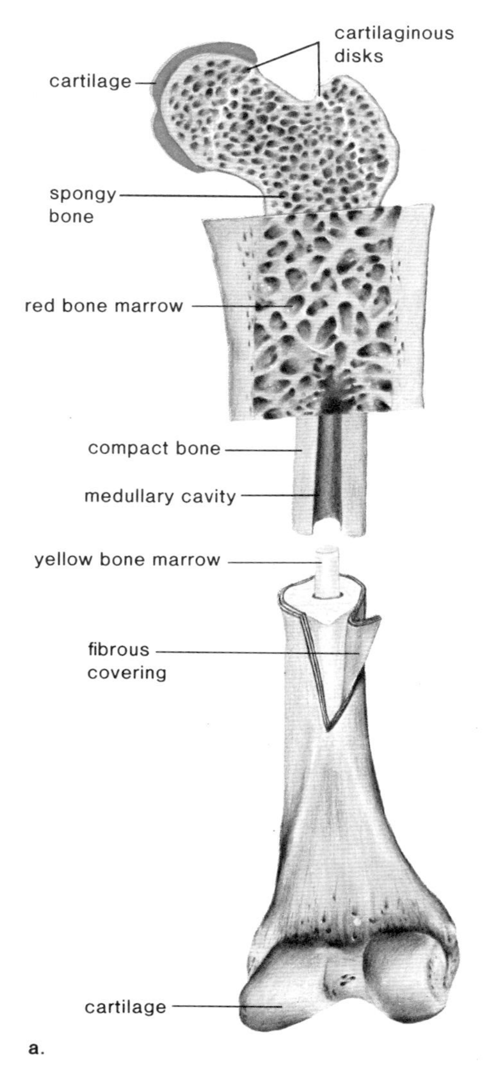

cartilaginous disks
cartilage
spongy bone
red bone marrow
compact bone
medullary cavity
yellow bone marrow
fibrous covering
cartilage
a.

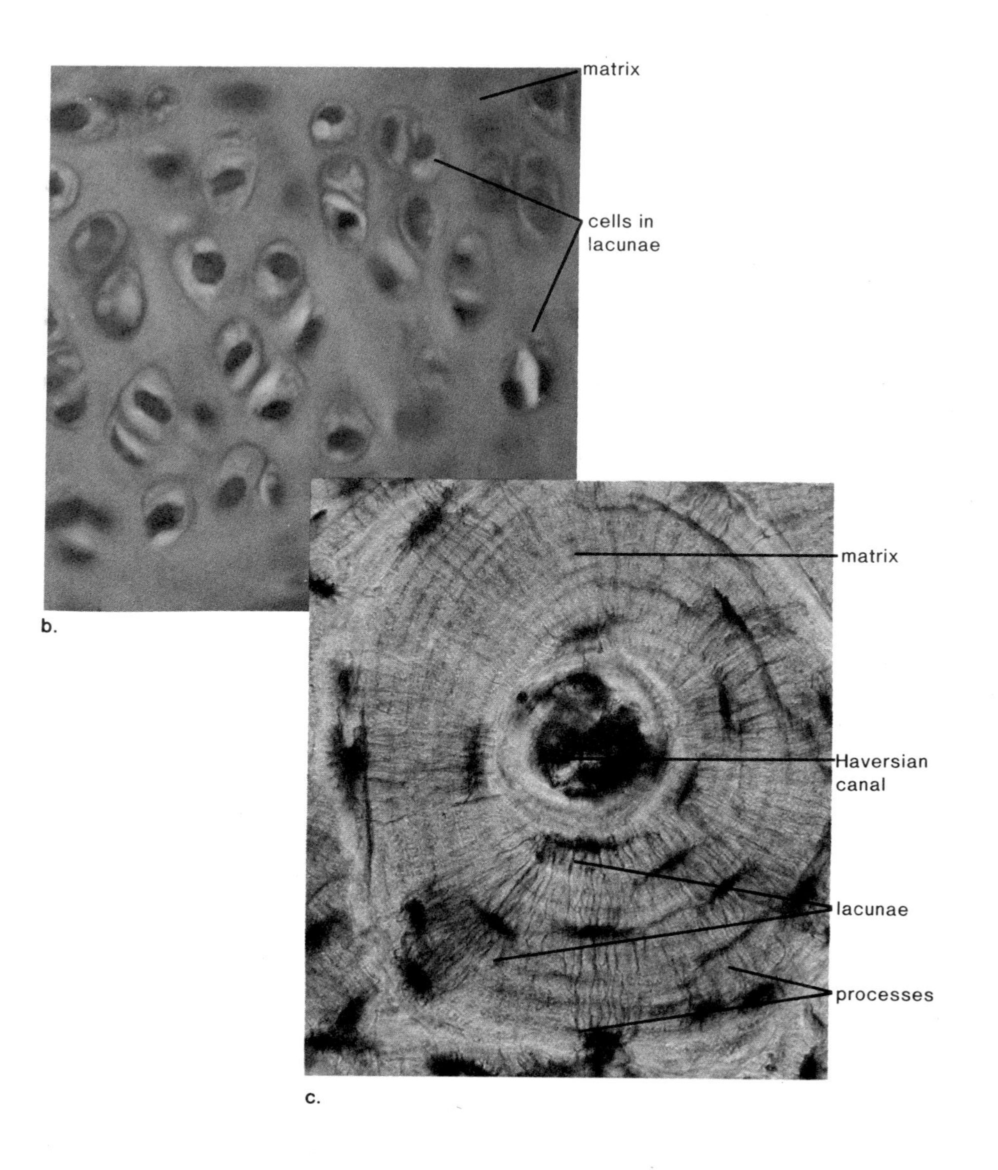

matrix
cells in lacunae
b.
matrix
Haversian canal
lacunae
processes
c.

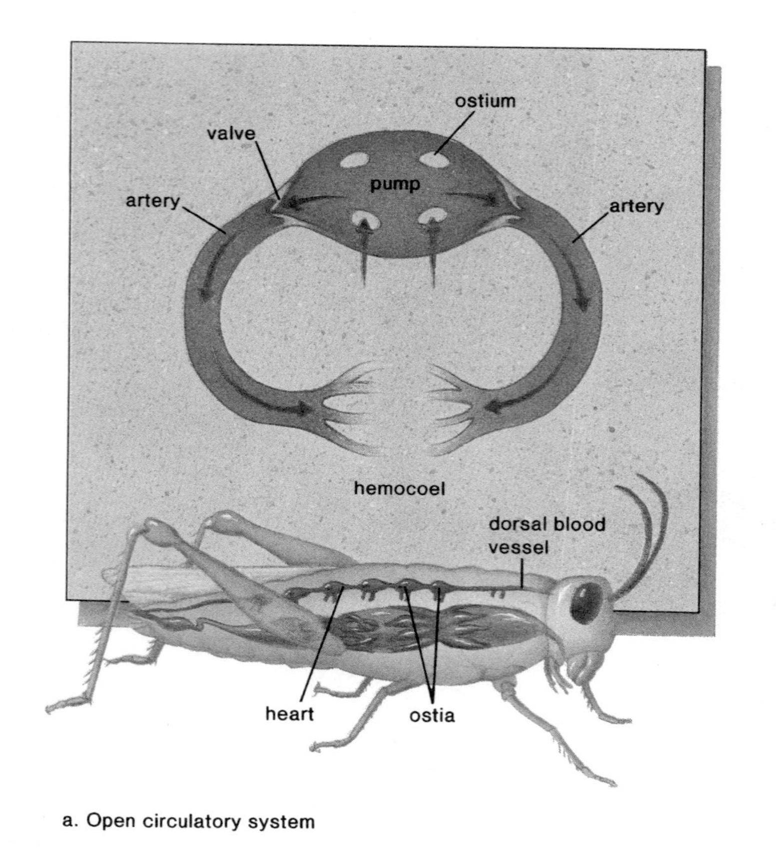
valve
ostium
pump
artery
artery
hemocoel
dorsal blood vessel
heart
ostia

a. Open circulatory system

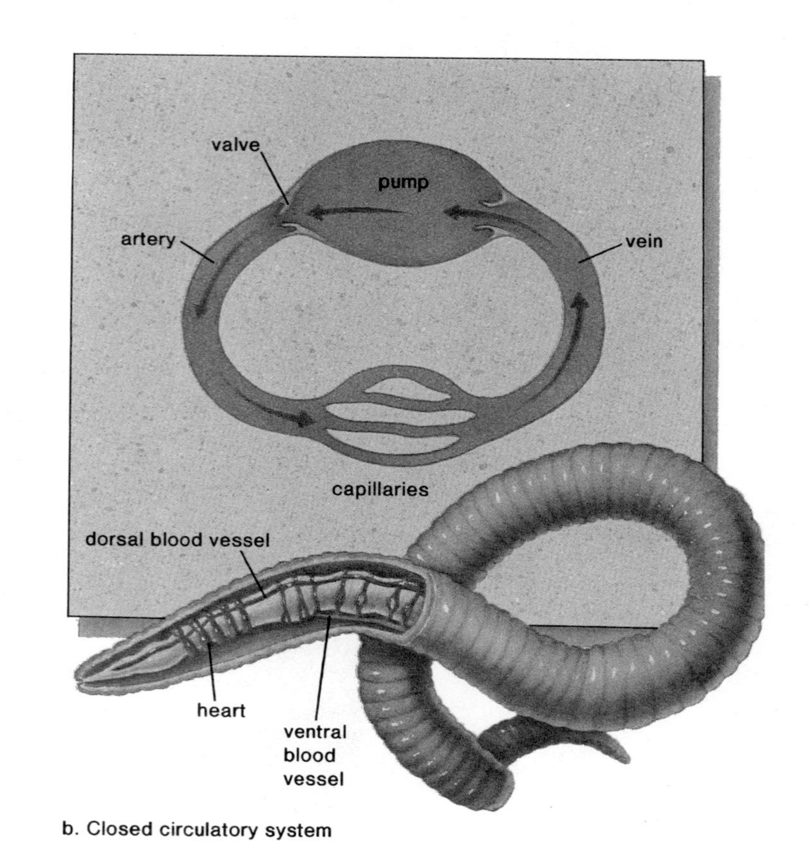
valve
pump
artery
vein
capillaries
dorsal blood vessel
heart
ventral blood vessel

b. Closed circulatory system

Open and Closed Circulatory Systems
Figure 16.1

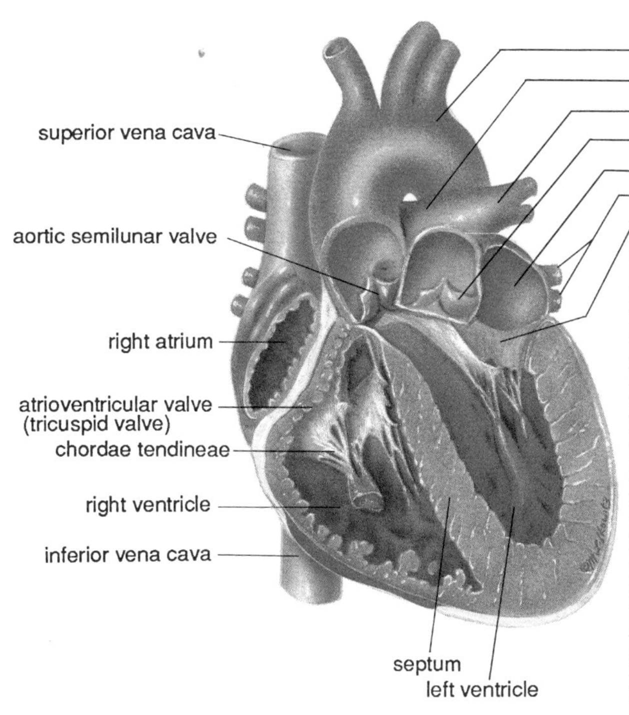
aorta
pulmonary trunk
left pulmonary artery
pulmonary semilunar valve
left atrium
left pulmonary veins
atrioventricular valve (bicuspid or mitral valve)
superior vena cava
aortic semilunar valve
right atrium
atrioventricular valve
(tricuspid valve)
chordae tendineae
right ventricle
inferior vena cava
septum
left ventricle

a. **Parts of the heart**

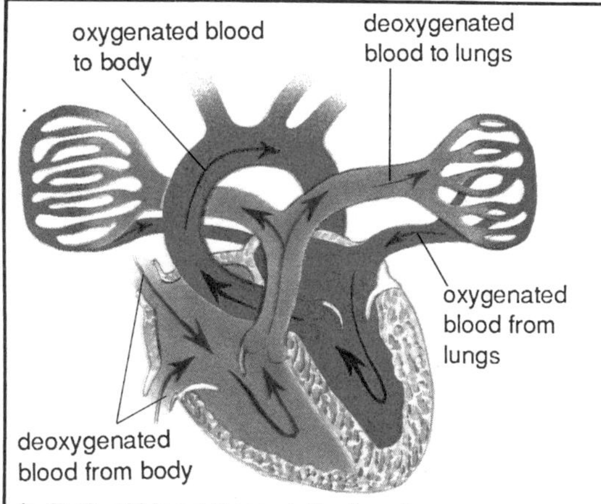
oxygenated blood
to body
deoxygenated
blood to lungs
oxygenated
blood from
lungs
deoxygenated
blood from body

b. **Path of blood through the heart**

- The superior (anterior) vena cava and the inferior (posterior) vena cava carrying deoxygenated blood (low in oxygen and high in carbon dioxide) enter the right atrium.
- The right atrium sends blood through an atrioventricular valve (the tricuspid valve) to the right ventricle.
- The right ventricle sends blood through the pulmonary semilunar valve into the pulmonary trunk and the pulmonary arteries to the lungs.
- The pulmonary veins carrying oxygenated blood (high in oxygen and low in carbon dioxide) from the lungs enter the left atrium.
- The left atrium sends blood through an atrioventricular valve (the bicuspid, or mitral, valve) to the left ventricle.
- The left ventricle sends blood through the aortic semilunar valve into the aorta to the body proper.

Internal Heart Anatomy
Figure 16.3

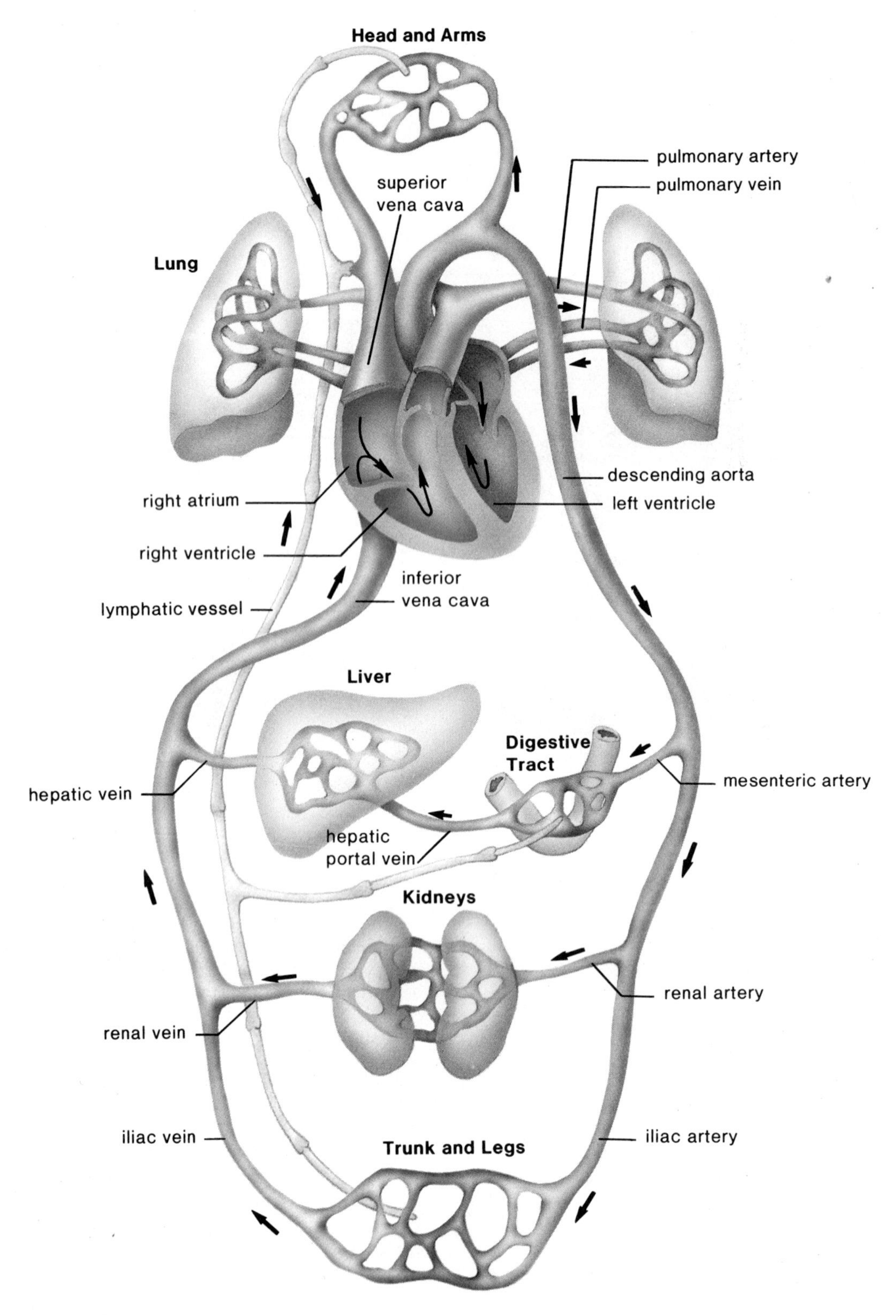
Head and Arms
superior vena cava
pulmonary artery
pulmonary vein
Lung
right atrium
right ventricle
lymphatic vessel
inferior vena cava
descending aorta
left ventricle
Liver
Digestive Tract
mesenteric artery
hepatic vein
hepatic portal vein
Kidneys
renal artery
renal vein
iliac vein
Trunk and Legs
iliac artery

Human Circulatory System
Figure 16.4

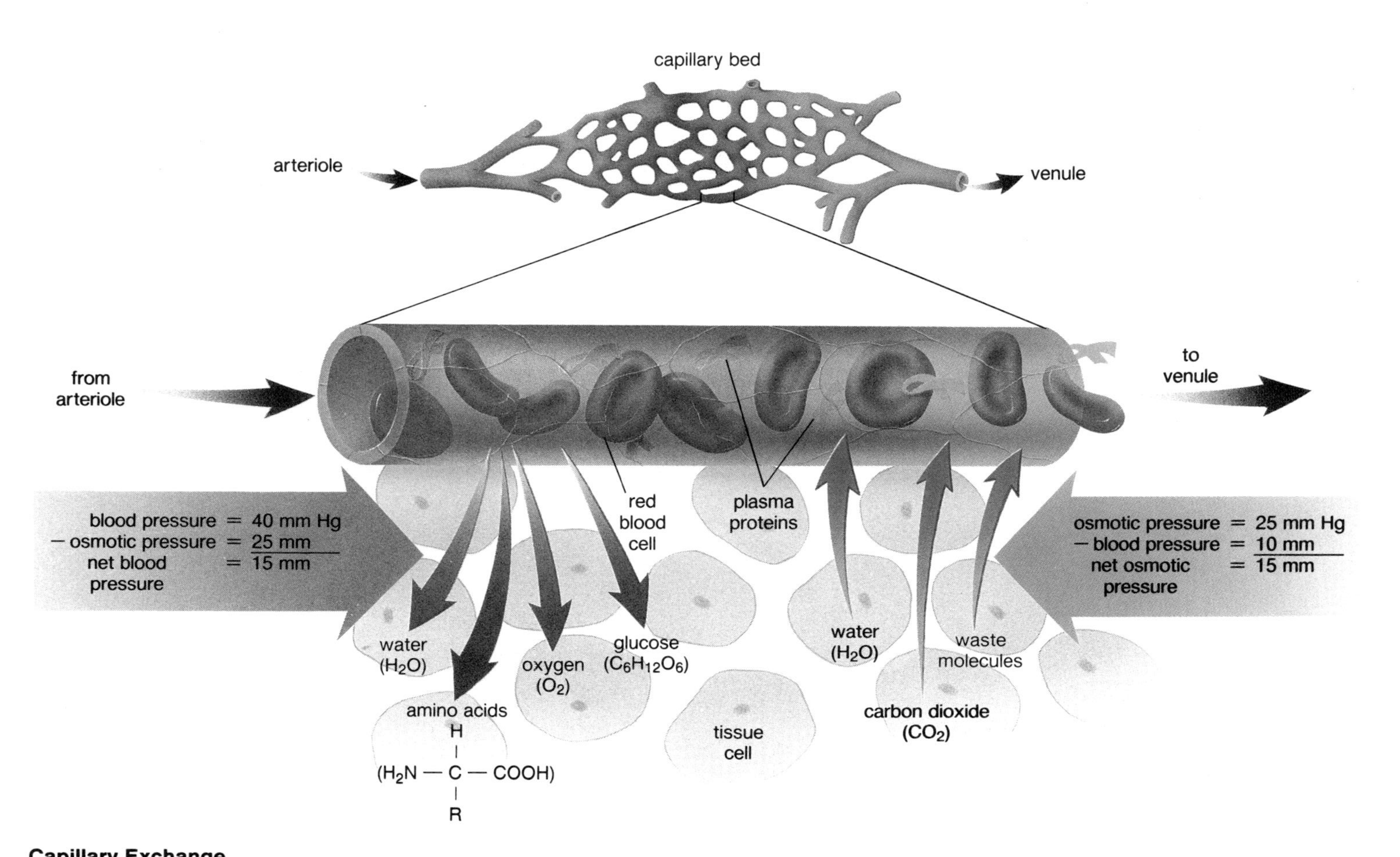

capillary bed
arteriole
venule
from arteriole
to venule
blood pressure = 40 mm Hg
− osmotic pressure = 25 mm
net blood pressure = 15 mm
red blood cell
plasma proteins
osmotic pressure = 25 mm Hg
− blood pressure = 10 mm
net osmotic pressure = 15 mm
water (H_2O)
amino acids
H
(H$_2$N — C — COOH)
R
oxygen (O_2)
glucose ($C_6H_{12}O_6$)
tissue cell
water (H_2O)
carbon dioxide (CO_2)
waste molecules

Capillary Exchange
Figure 16.9

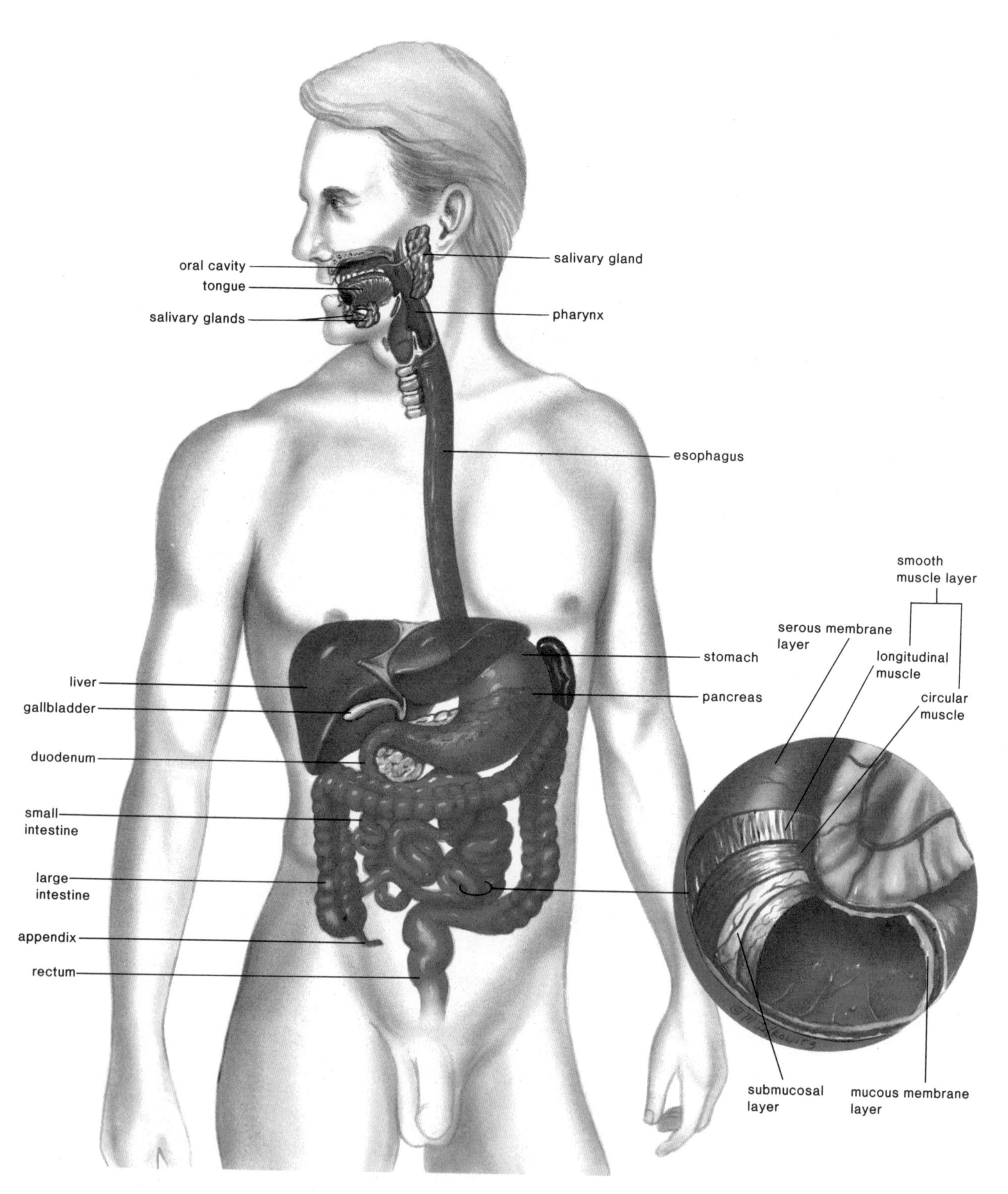
oral cavity
tongue
salivary glands
salivary gland
pharynx
esophagus
liver
gallbladder
duodenum
small intestine
large intestine
appendix
rectum
stomach
pancreas
smooth muscle layer
serous membrane layer
longitudinal muscle
circular muscle
submucosal layer
mucous membrane layer

Human Digestive System
Figure 17.4

Anatomy of Intestinal Villus
Figure 17.5

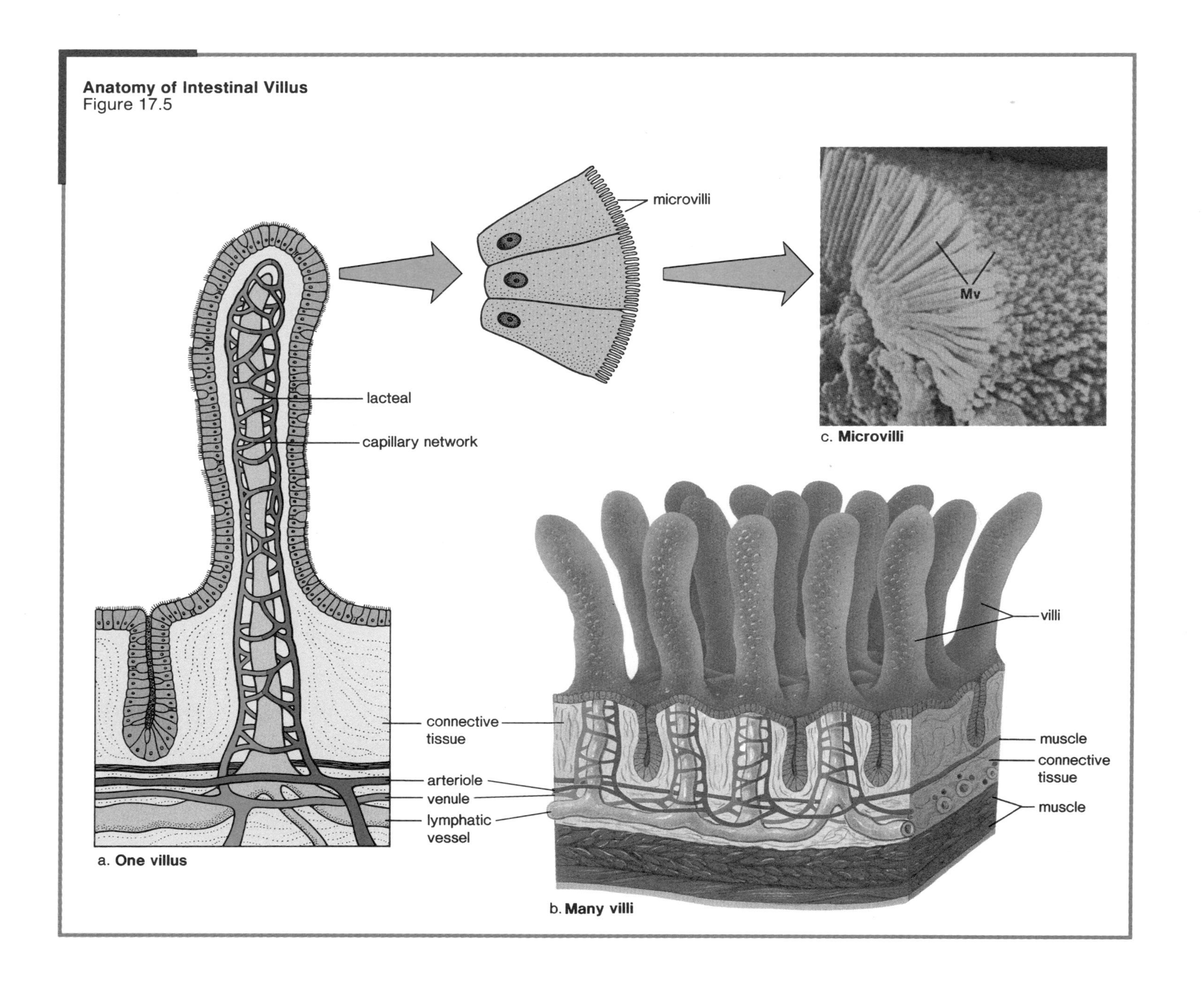

lacteal
capillary network
microvilli
Mv
connective tissue
arteriole
venule
lymphatic vessel
villi
muscle
connective tissue
muscle

a. **One villus**

b. **Many villi**

c. **Microvilli**

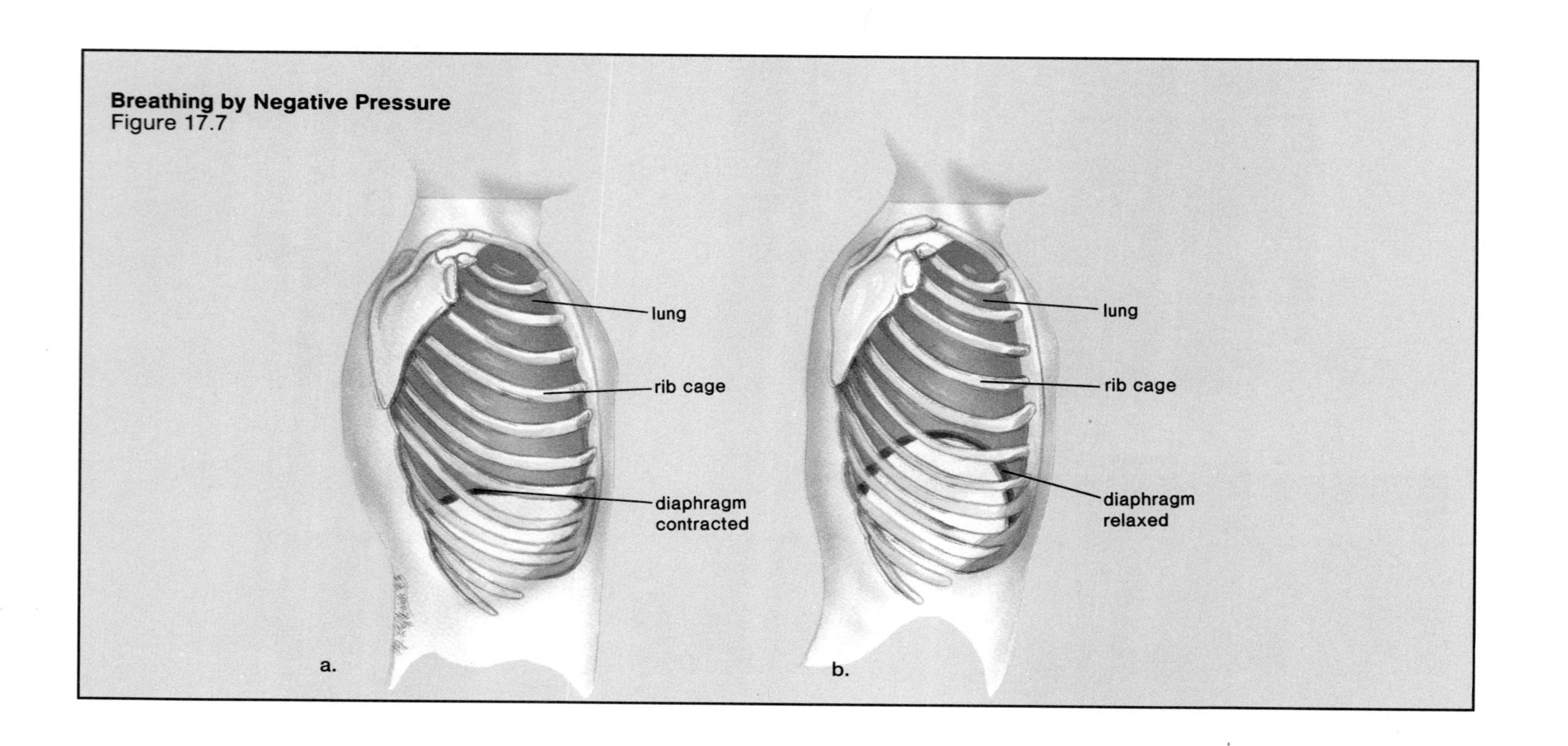
Breathing by Negative Pressure
Figure 17.7
lung
rib cage
diaphragm contracted
a.
lung
rib cage
diaphragm relaxed
b.

Human Respiratory System
Figure 17.9

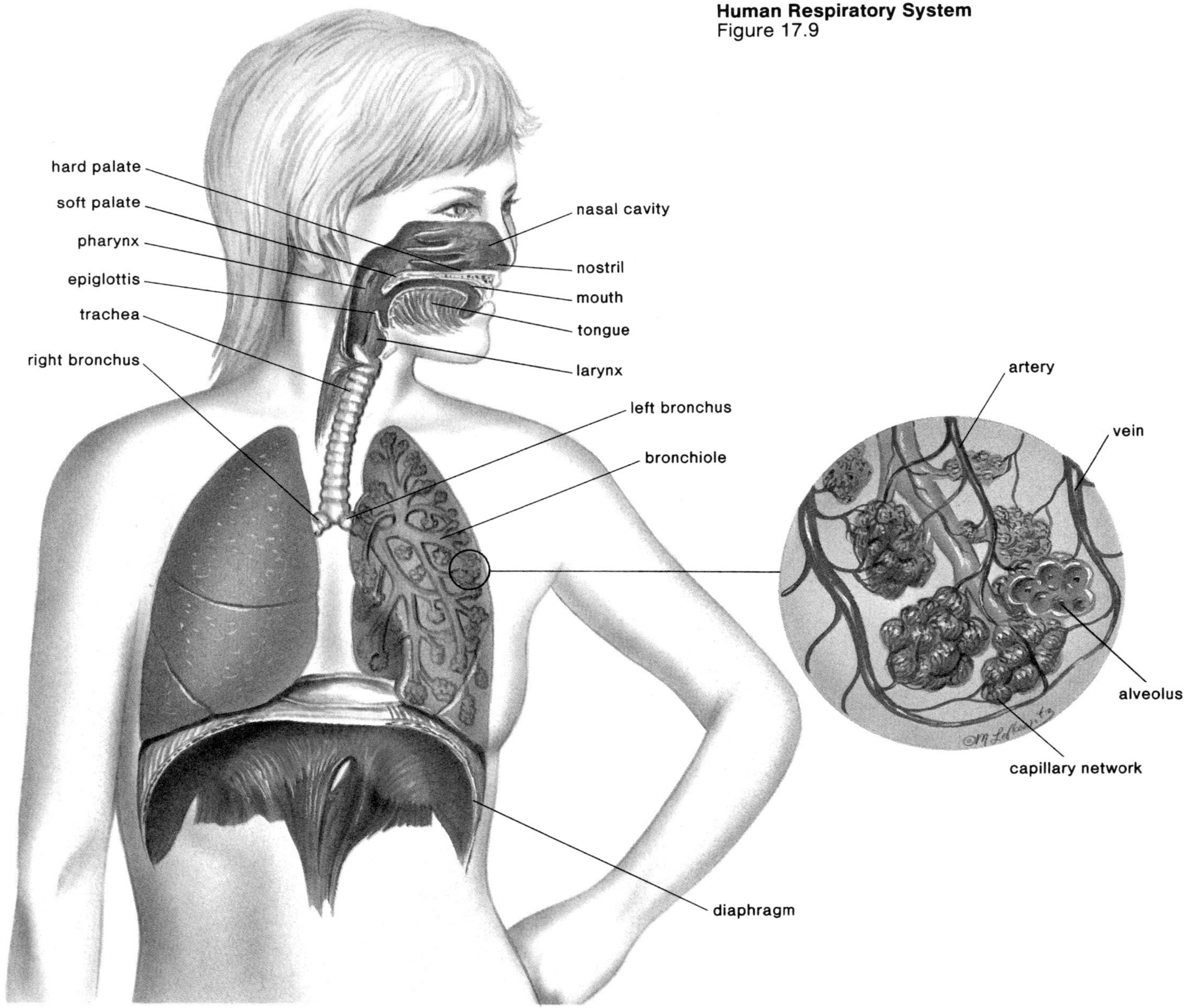

hard palate
soft palate
pharynx
epiglottis
trachea
right bronchus
nasal cavity
nostril
mouth
tongue
larynx
left bronchus
bronchiole
diaphragm
artery
vein
alveolus
capillary network

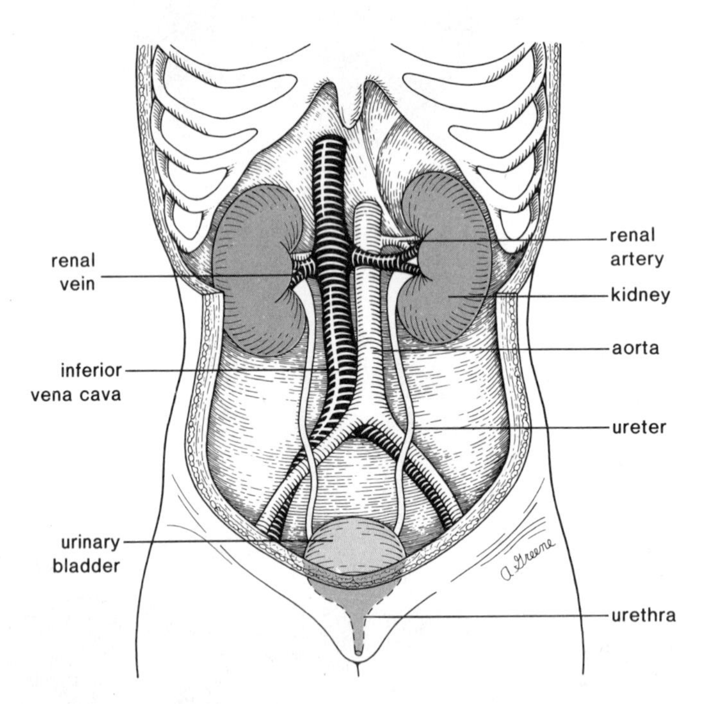
renal vein
inferior vena cava
urinary bladder
renal artery
kidney
aorta
ureter
urethra

The Urinary System
Figure 17.12

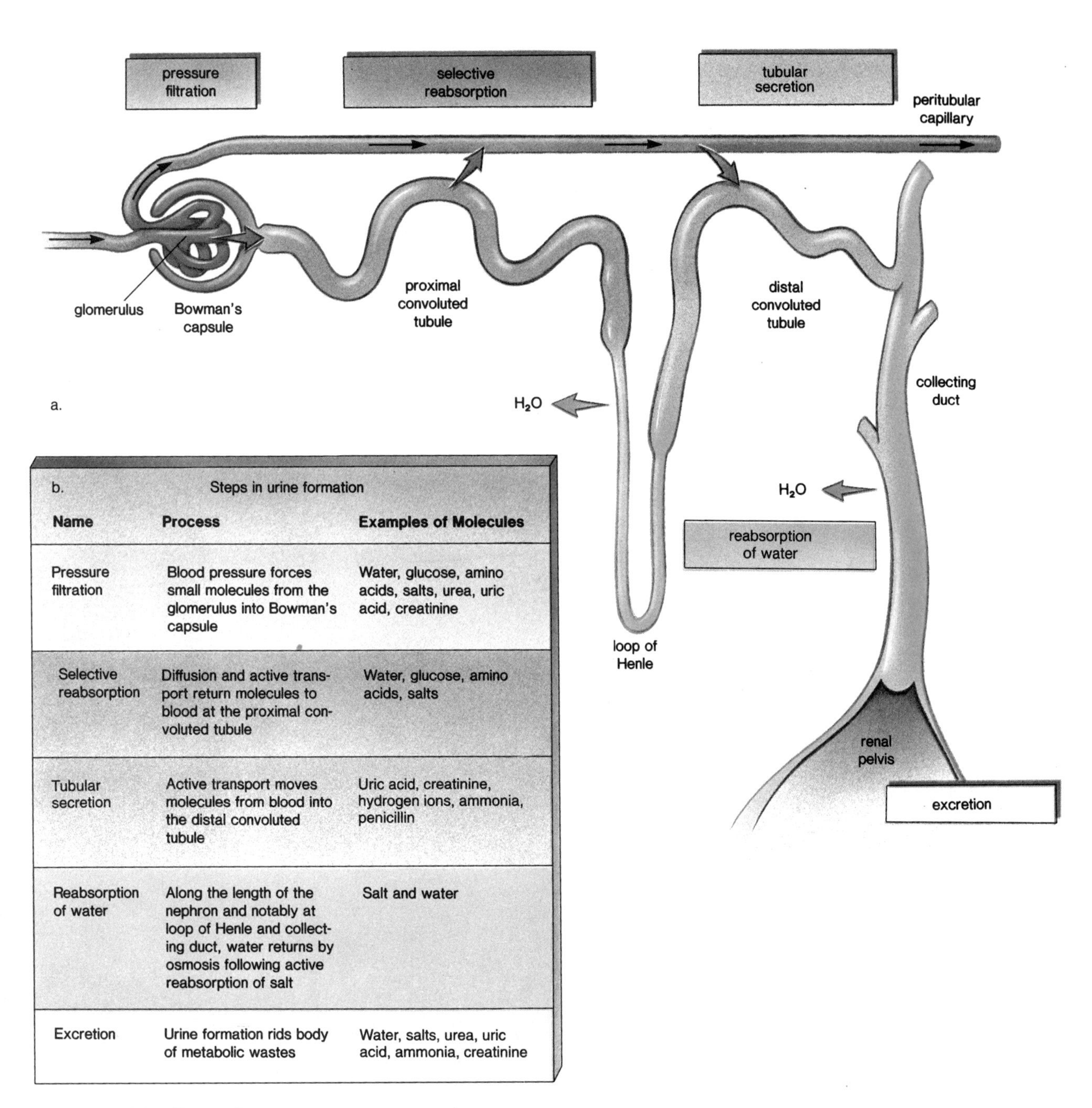
pressure
filtration
selective
reabsorption
tubular
secretion
peritubular
capillary
glomerulus
Bowman's
capsule
proximal
convoluted
tubule
distal
convoluted
tubule
collecting
duct
a.
H_2O
H_2O
reabsorption
of water
loop of
Henle
renal
pelvis
excretion

b. Steps in urine formation

Name	Process	Examples of Molecules
Pressure filtration	Blood pressure forces small molecules from the glomerulus into Bowman's capsule	Water, glucose, amino acids, salts, urea, uric acid, creatinine
Selective reabsorption	Diffusion and active transport return molecules to blood at the proximal convoluted tubule	Water, glucose, amino acids, salts
Tubular secretion	Active transport moves molecules from blood into the distal convoluted tubule	Uric acid, creatinine, hydrogen ions, ammonia, penicillin
Reabsorption of water	Along the length of the nephron and notably at loop of Henle and collecting duct, water returns by osmosis following active reabsorption of salt	Salt and water
Excretion	Urine formation rids body of metabolic wastes	Water, salts, urea, uric acid, ammonia, creatinine

Steps in Urine Formation
Figure 17.14

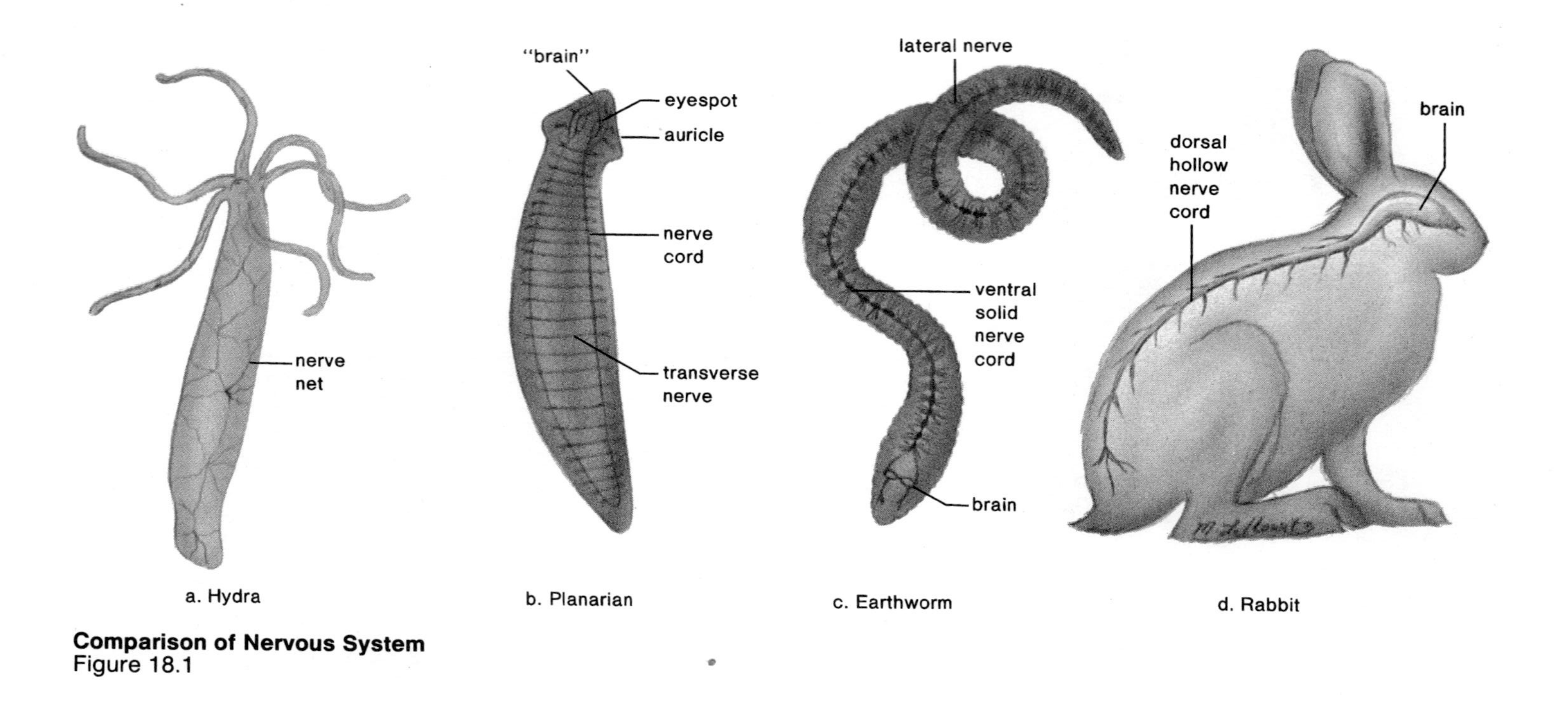
nerve
net
"brain"
eyespot
auricle
nerve
cord
transverse
nerve
lateral nerve
ventral
solid
nerve
cord
brain
dorsal
hollow
nerve
cord
brain
a. Hydra
b. Planarian
c. Earthworm
d. Rabbit

Comparison of Nervous System
Figure 18.1

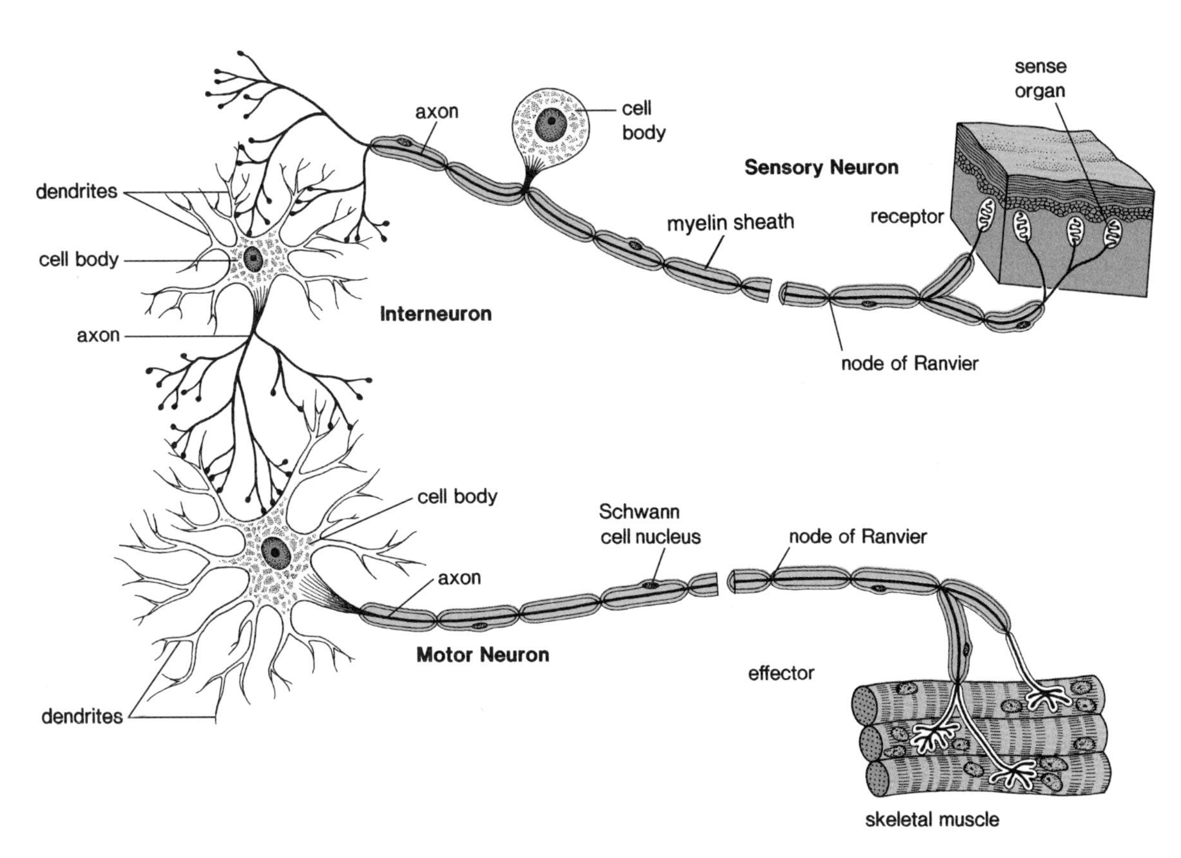
sense
organ
axon
cell
body
Sensory Neuron
dendrites
myelin sheath
receptor
cell body
Interneuron
axon
node of Ranvier
cell body
Schwann
cell nucleus
node of Ranvier
axon
Motor Neuron
effector
dendrites
skeletal muscle

Types of Neurons
Figure 18.2

Synapse Structure and Function
Figure 18.4

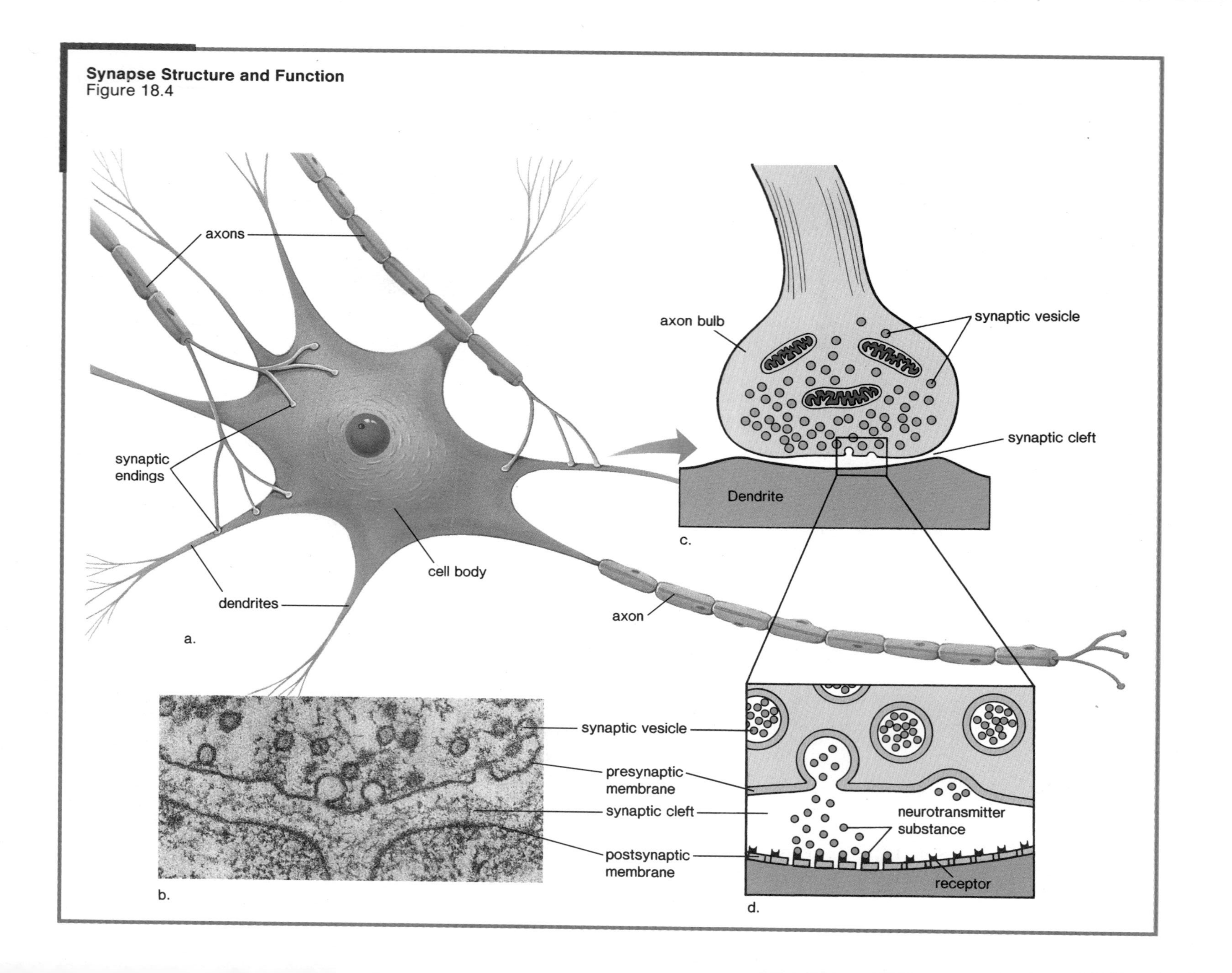
axons
synaptic
endings
dendrites
cell body
a.
axon
axon bulb
synaptic vesicle
synaptic cleft
Dendrite
c.
b.
synaptic vesicle
presynaptic
membrane
synaptic cleft
postsynaptic
membrane
neurotransmitter
substance
receptor
d.

Central Nervous System

brain

spinal cord

Peripheral Nervous System

cranial nerves

spinal nerves

somatic system (to skeletal muscles)

autonomic system (to smooth muscles)

sympathetic system

parasympathetic system

b.

Organization of Human Nervous System
Figure 18.5b

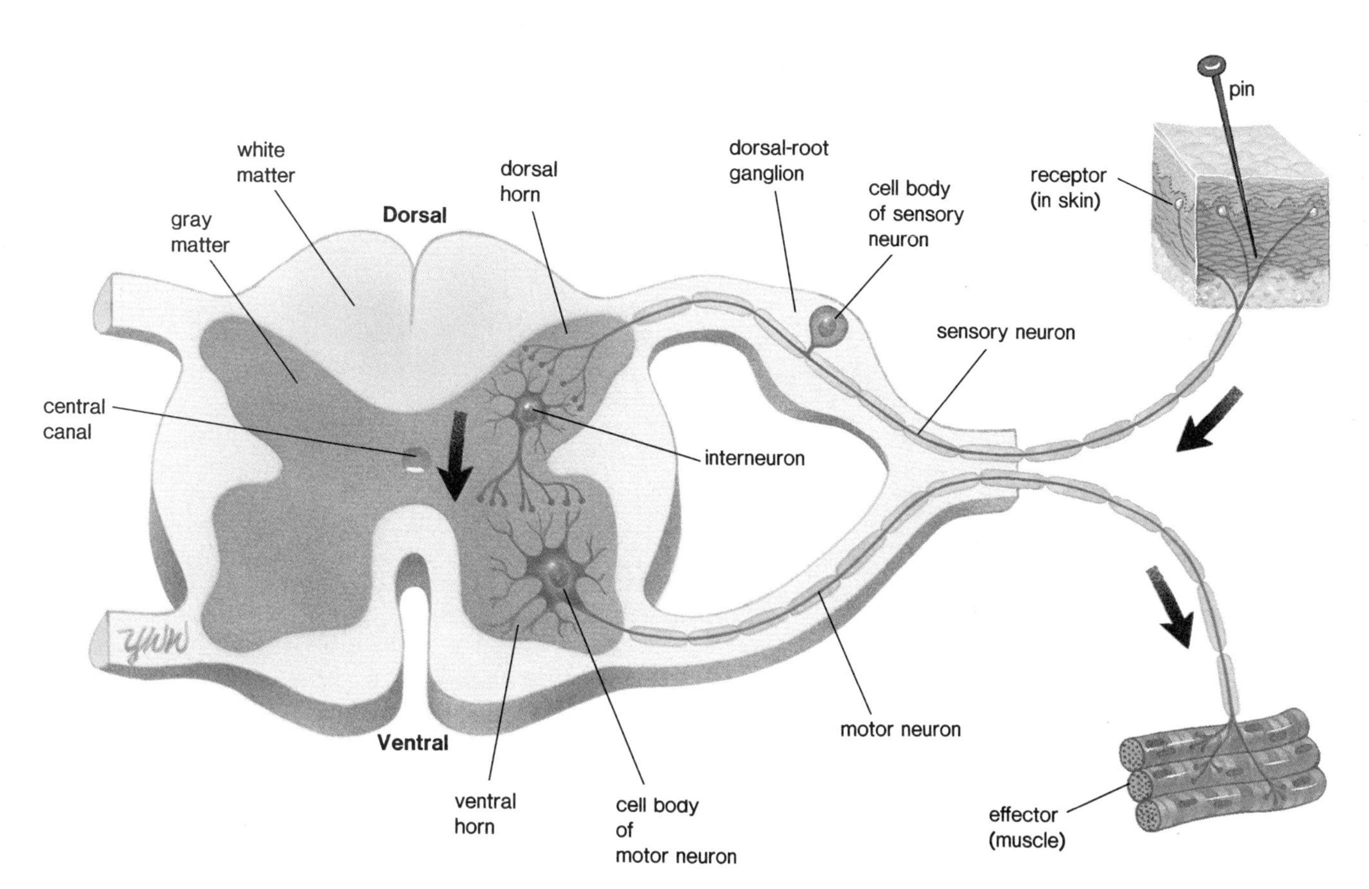
pin
white matter
dorsal horn
dorsal-root ganglion
cell body of sensory neuron
receptor (in skin)
gray matter
Dorsal
sensory neuron
central canal
interneuron
motor neuron
Ventral
ventral horn
cell body of motor neuron
effector (muscle)

Spinal Cord and Reflex Arc
Figure 18.6

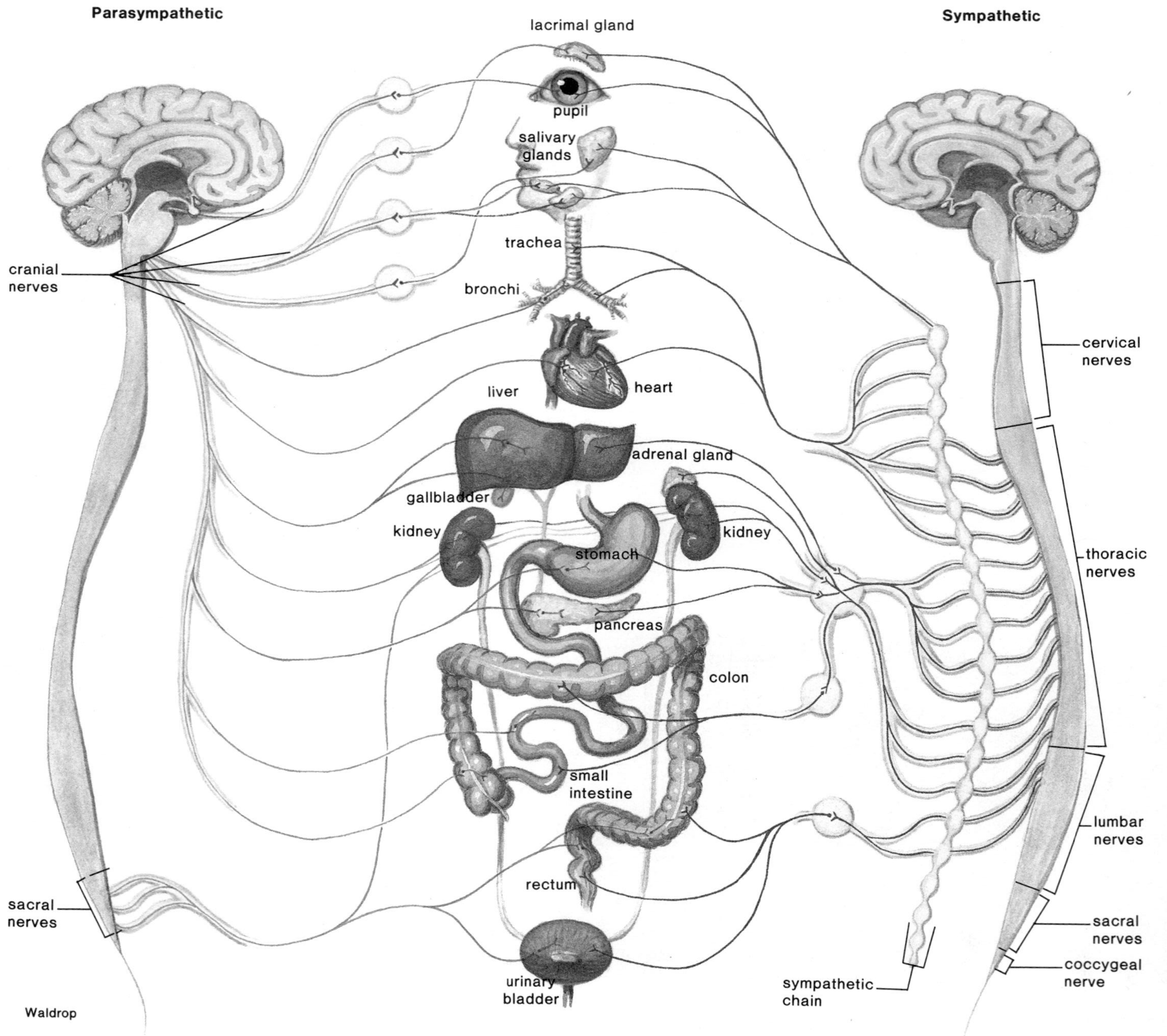

Parasympathetic

Sympathetic

lacrimal gland
pupil
salivary glands
trachea
bronchi
heart
liver
adrenal gland
gallbladder
kidney
kidney
stomach
pancreas
colon
small intestine
rectum
urinary bladder
cranial nerves
sacral nerves
cervical nerves
thoracic nerves
lumbar nerves
sacral nerves
coccygeal nerve
sympathetic chain

Waldrop

Autonomic Nervous System
Figure 18.7

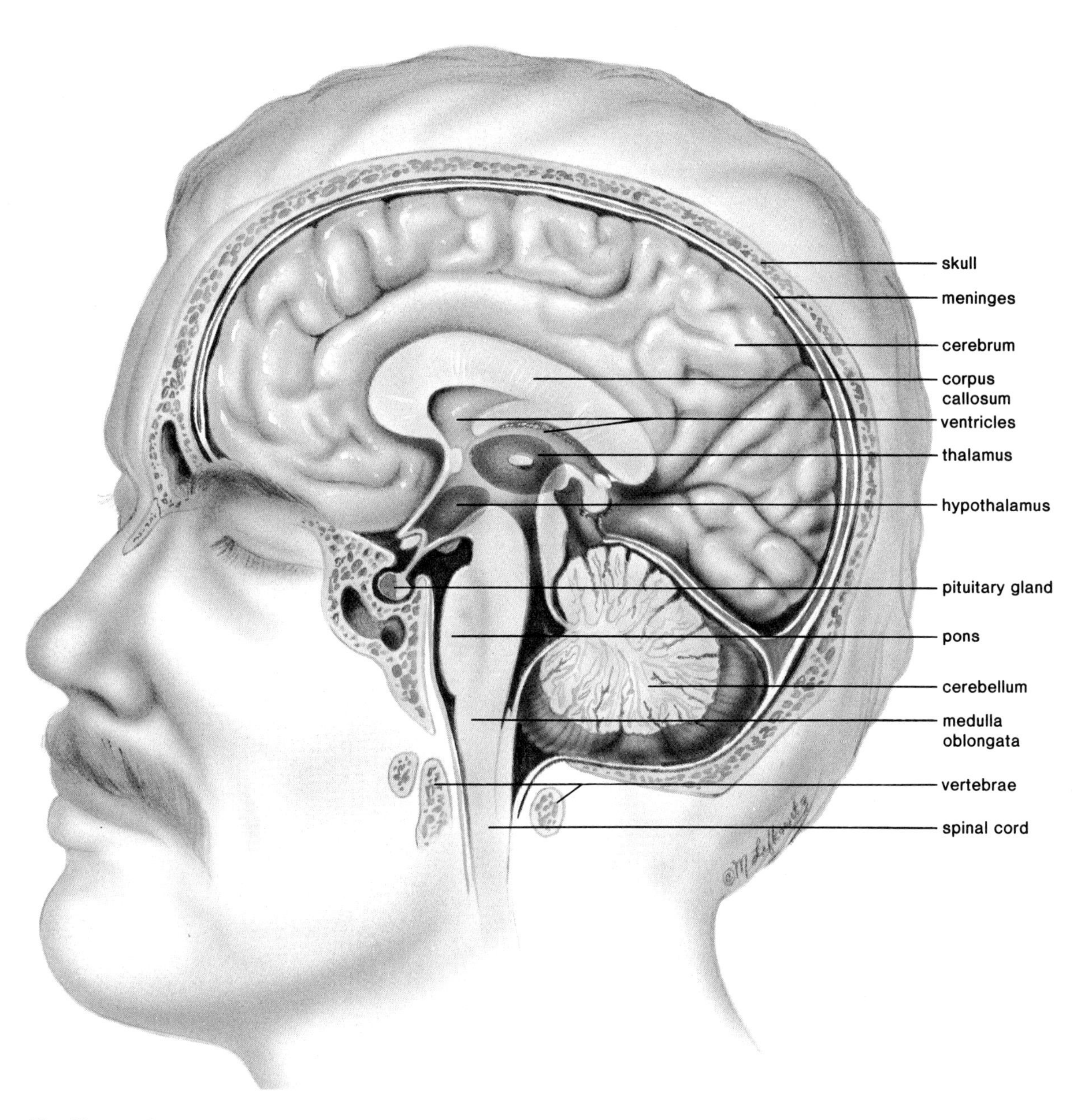
skull
meninges
cerebrum
corpus
callosum
ventricles
thalamus
hypothalamus
pituitary gland
pons
cerebellum
medulla
oblongata
vertebrae
spinal cord

The Human Brain
Figure 18.8

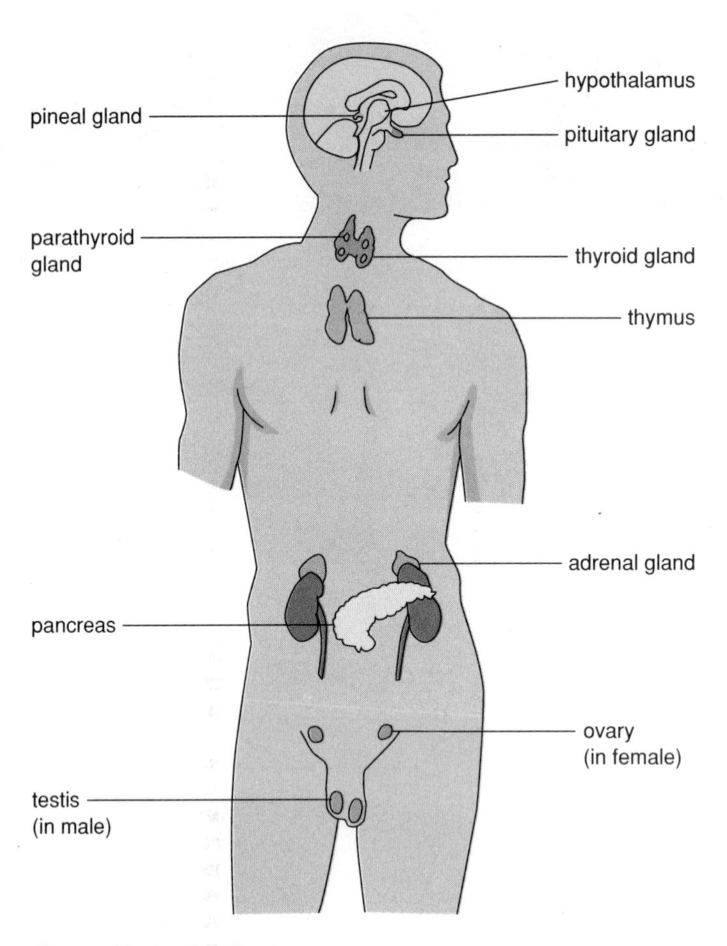
hypothalamus
pineal gland
pituitary gland
parathyroid gland
thyroid gland
thymus
adrenal gland
pancreas
ovary (in female)
testis (in male)

Human Endocrine System
Figure 18.10

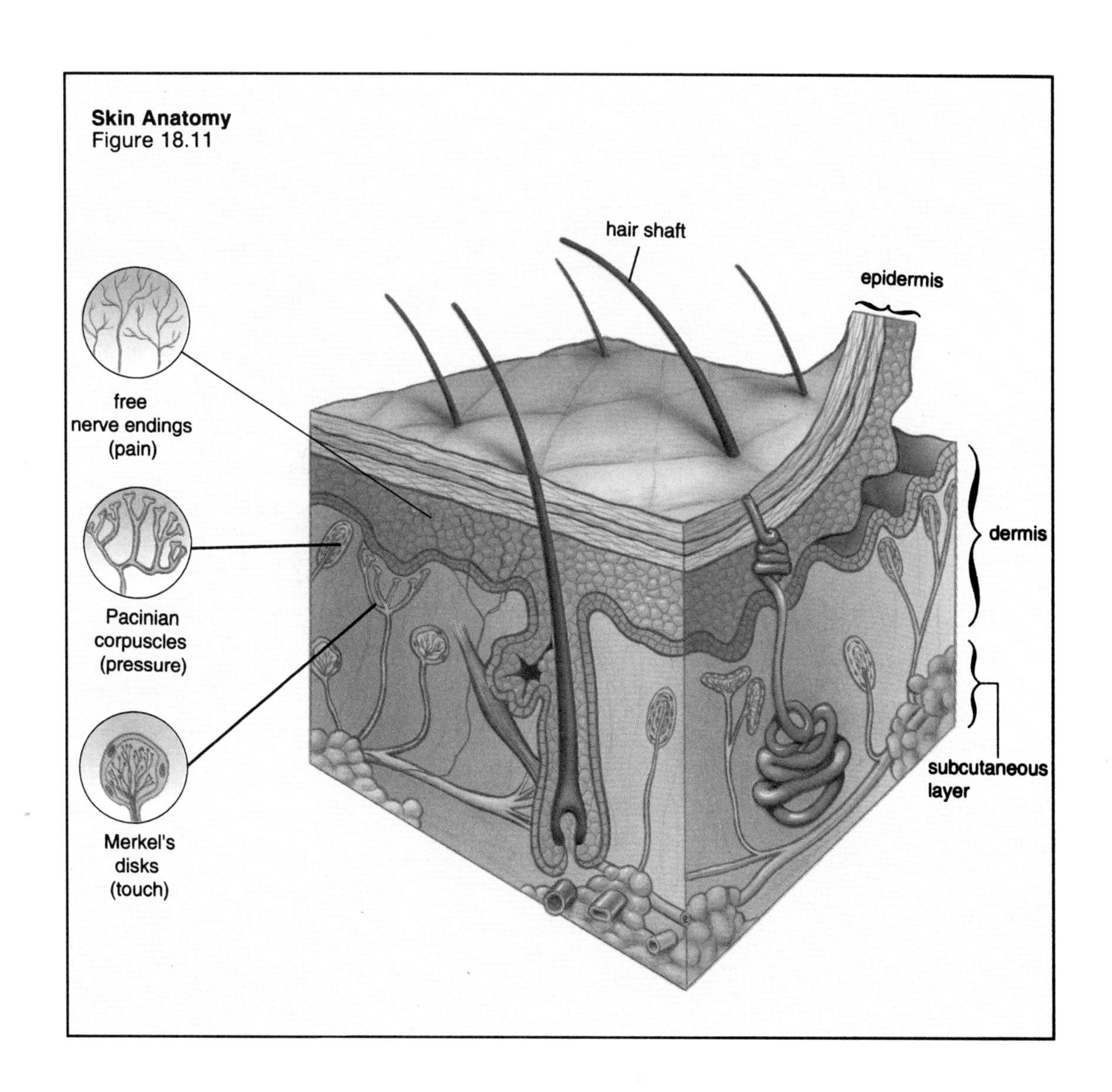
Skin Anatomy
Figure 18.11
hair shaft
epidermis
dermis
subcutaneous layer
free nerve endings (pain)
Pacinian corpuscles (pressure)
Merkel's disks (touch)

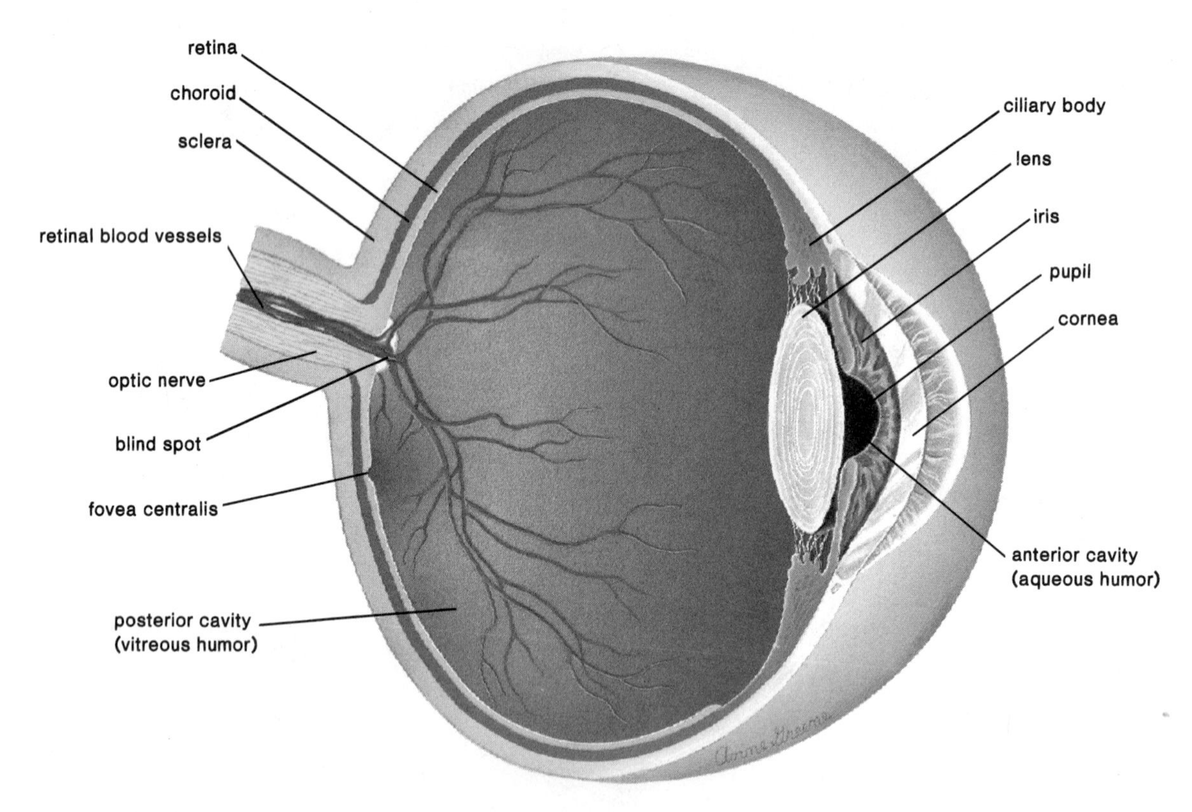
retina
choroid
sclera
retinal blood vessels
optic nerve
blind spot
fovea centralis
posterior cavity
(vitreous humor)
ciliary body
lens
iris
pupil
cornea
anterior cavity
(aqueous humor)

Human Eye Anatomy
Figure 18.12

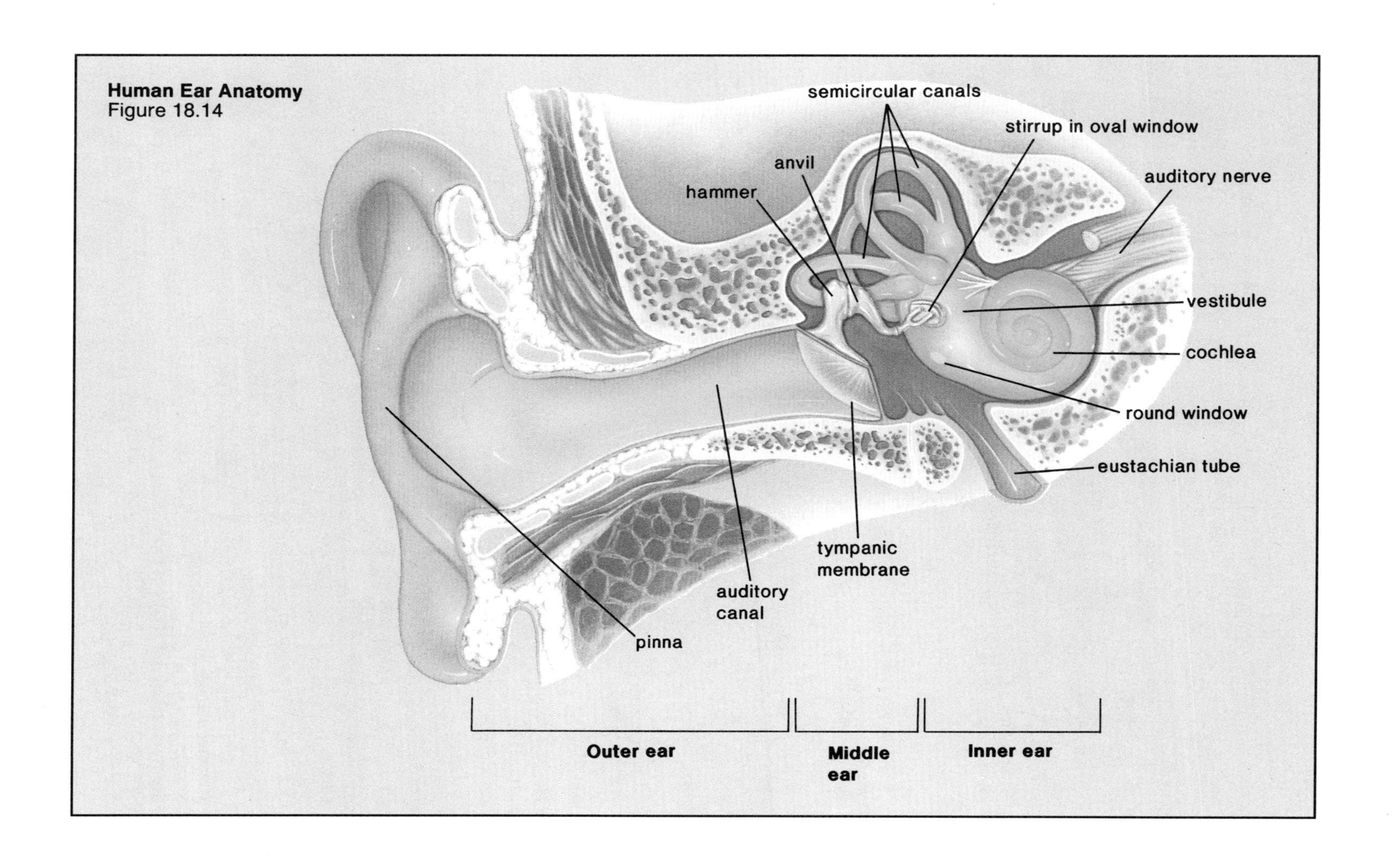
semicircular canals
stirrup in oval window
anvil
hammer
auditory nerve
vestibule
cochlea
round window
eustachian tube
tympanic membrane
auditory canal
pinna
Outer ear
Middle ear
Inner ear

Human Ear Anatomy
Figure 18.14

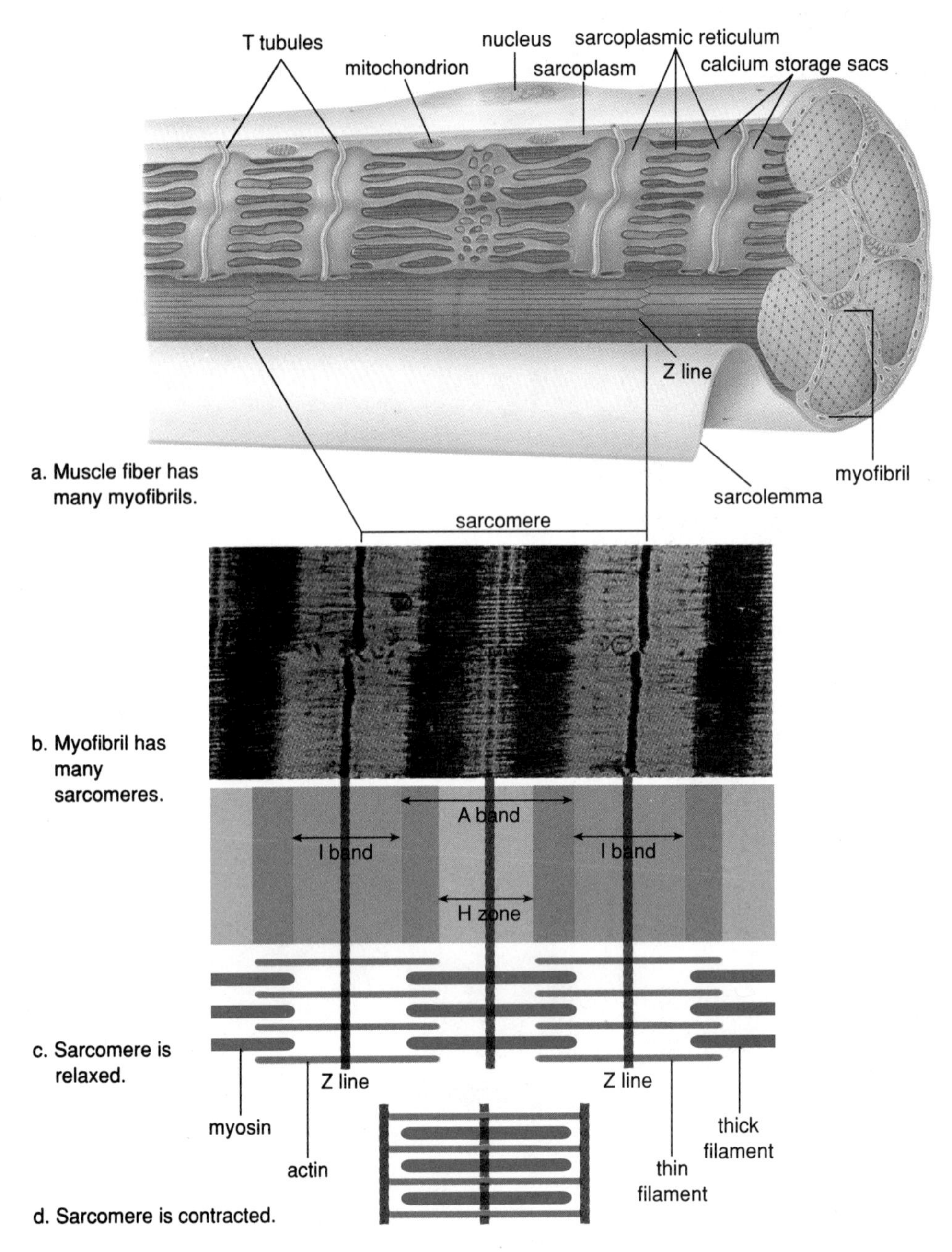
T tubules
mitochondrion
nucleus
sarcoplasm
sarcoplasmic reticulum
calcium storage sacs
Z line
myofibril
sarcolemma
a. Muscle fiber has many myofibrils.
sarcomere
b. Myofibril has many sarcomeres.
A band
I band
I band
H zone
c. Sarcomere is relaxed.
Z line
Z line
myosin
actin
thick filament
thin filament
d. Sarcomere is contracted.

Anatomy of Skeletal Muscle Fiber
Figure 18.15

Male Reproductive System
Figure 19.1

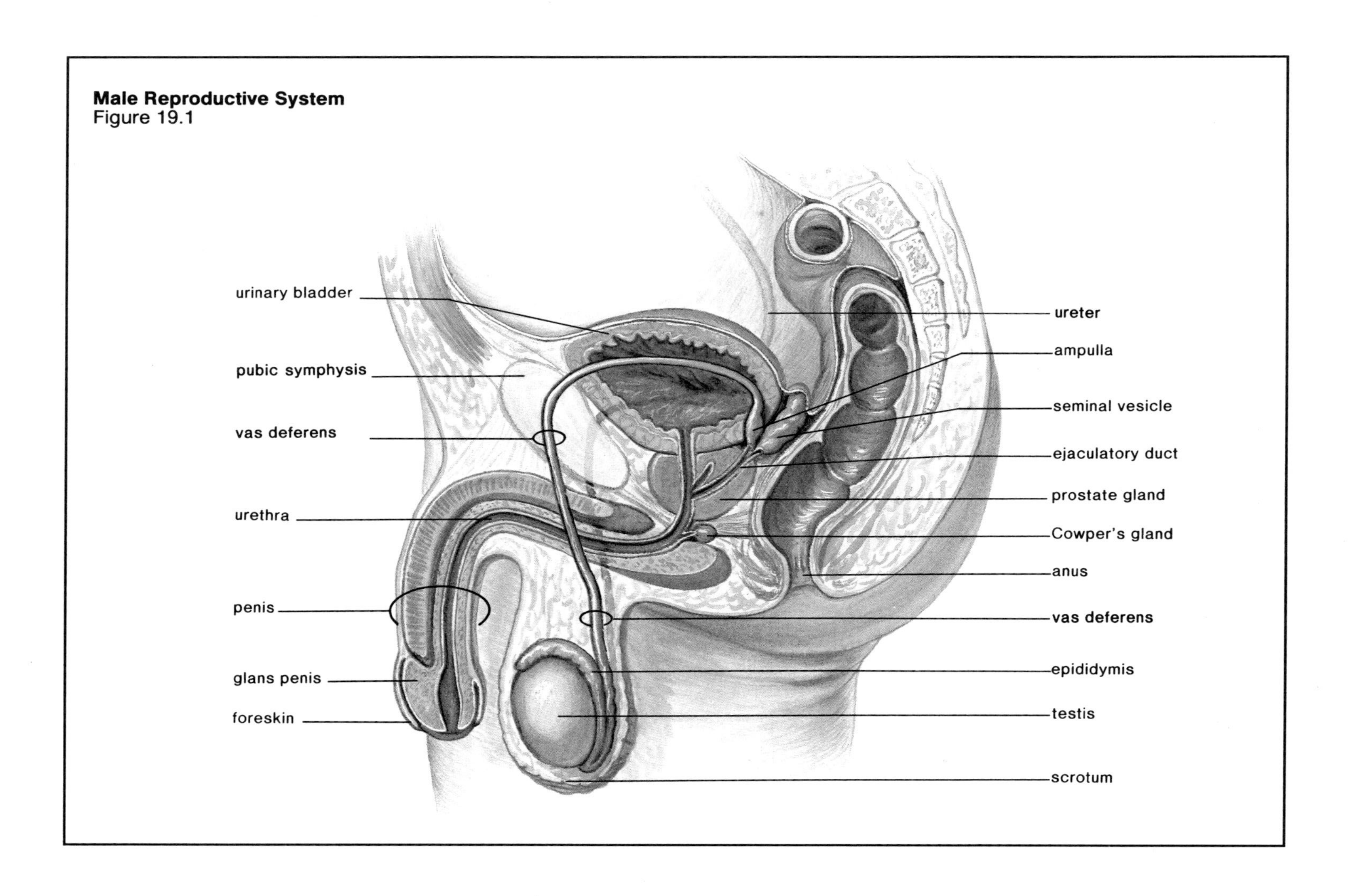
urinary bladder
pubic symphysis
vas deferens
urethra
penis
glans penis
foreskin
ureter
ampulla
seminal vesicle
ejaculatory duct
prostate gland
Cowper's gland
anus
vas deferens
epididymis
testis
scrotum

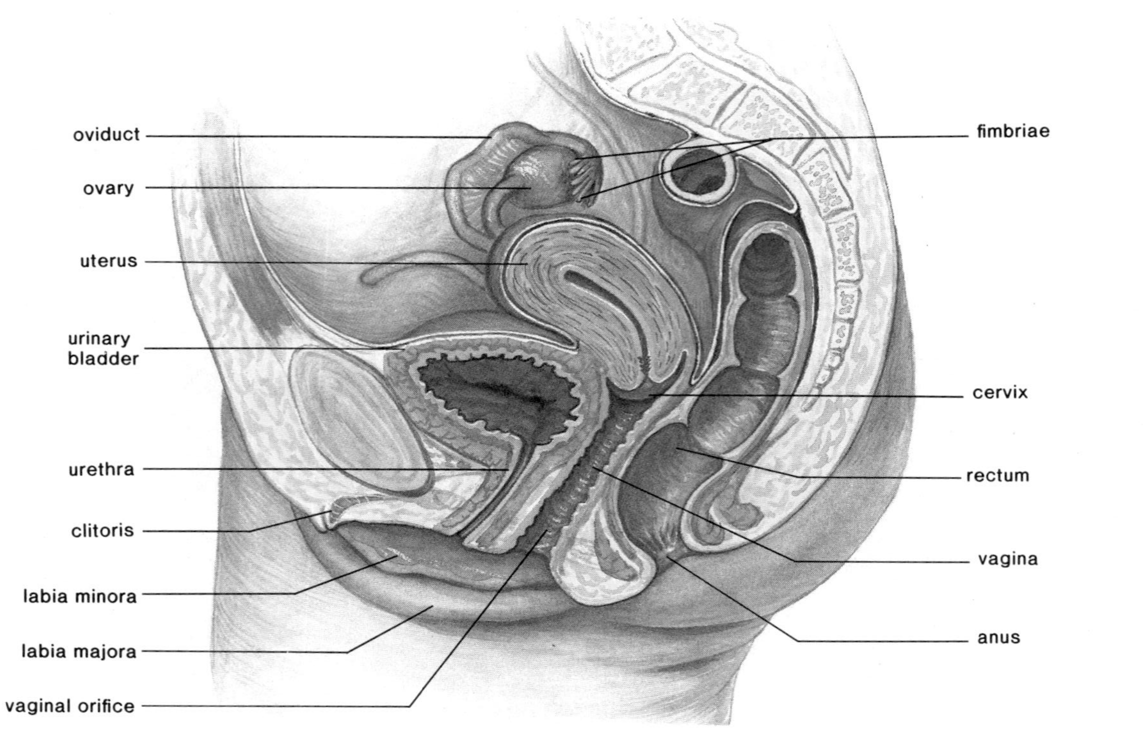
oviduct
ovary
uterus
urinary bladder
urethra
clitoris
labia minora
labia majora
vaginal orifice
fimbriae
cervix
rectum
vagina
anus

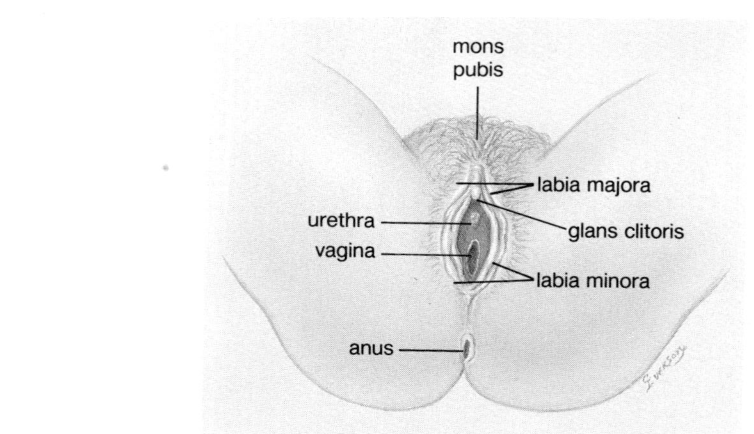
mons pubis
labia majora
urethra
glans clitoris
vagina
labia minora
anus

Female Reproductive System
Figure 19.3

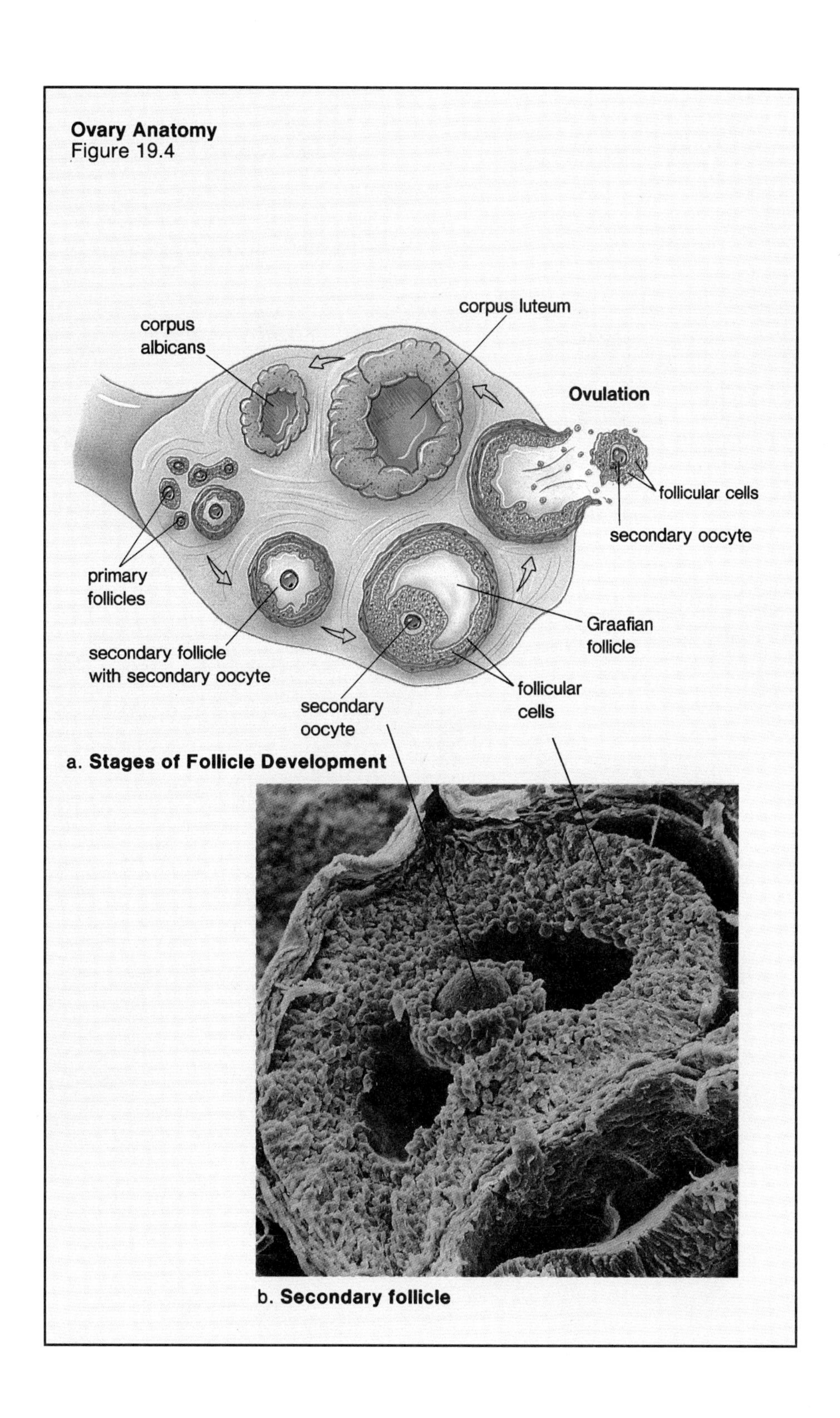
Ovary Anatomy
Figure 19.4
corpus luteum
corpus
albicans
Ovulation
follicular cells
secondary oocyte
primary
follicles
Graafian
follicle
secondary follicle
with secondary oocyte
follicular
cells
secondary
oocyte

a. **Stages of Follicle Development**

b. **Secondary follicle**

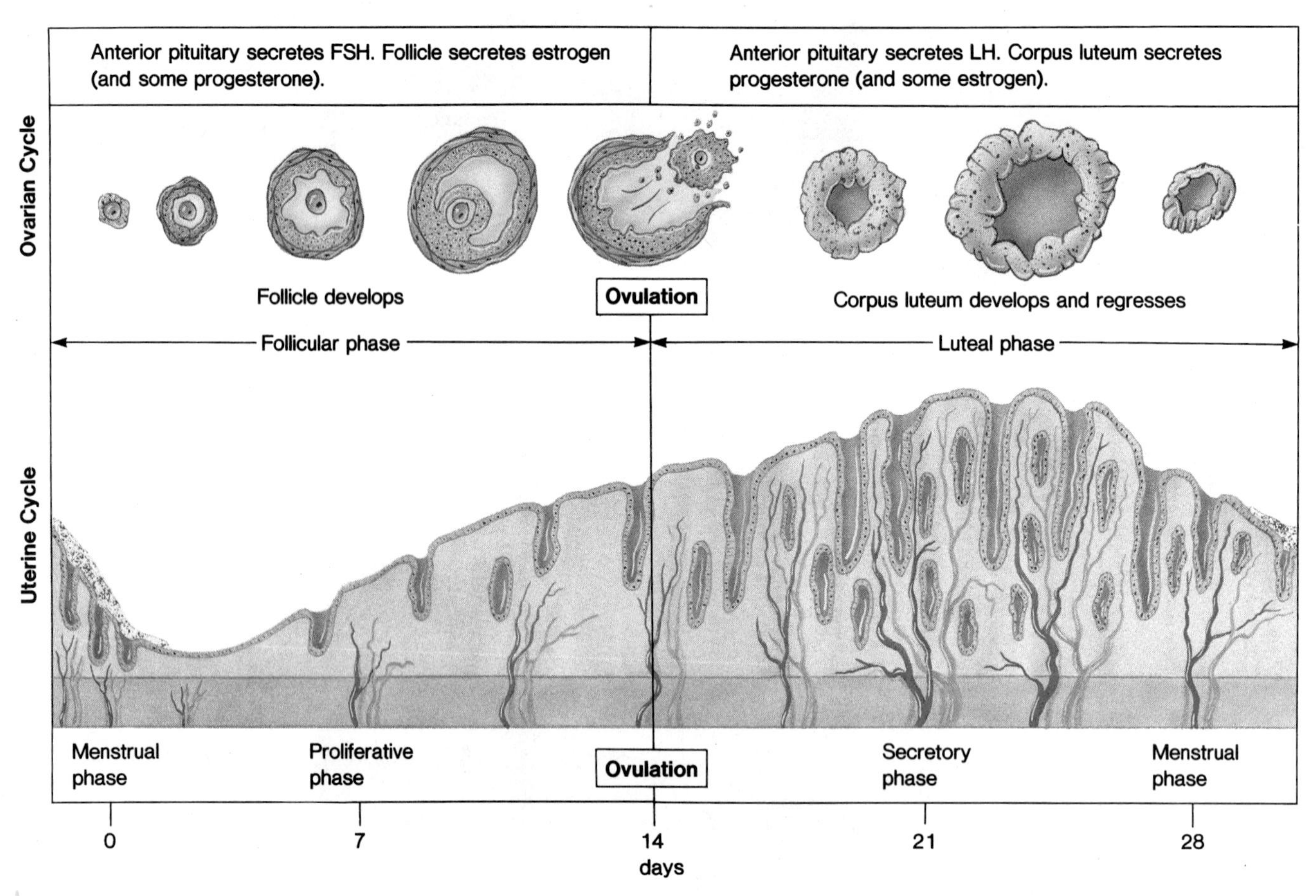

Anterior pituitary secretes FSH. Follicle secretes estrogen (and some progesterone).
Anterior pituitary secretes LH. Corpus luteum secretes progesterone (and some estrogen).
Ovarian Cycle
Follicle develops
Ovulation
Corpus luteum develops and regresses
Follicular phase
Luteal phase
Uterine Cycle
Menstrual phase
Proliferative phase
Ovulation
Secretory phase
Menstrual phase
0
7
14
days
21
28

Ovarian and Uterine Cycles
Figure 19.6

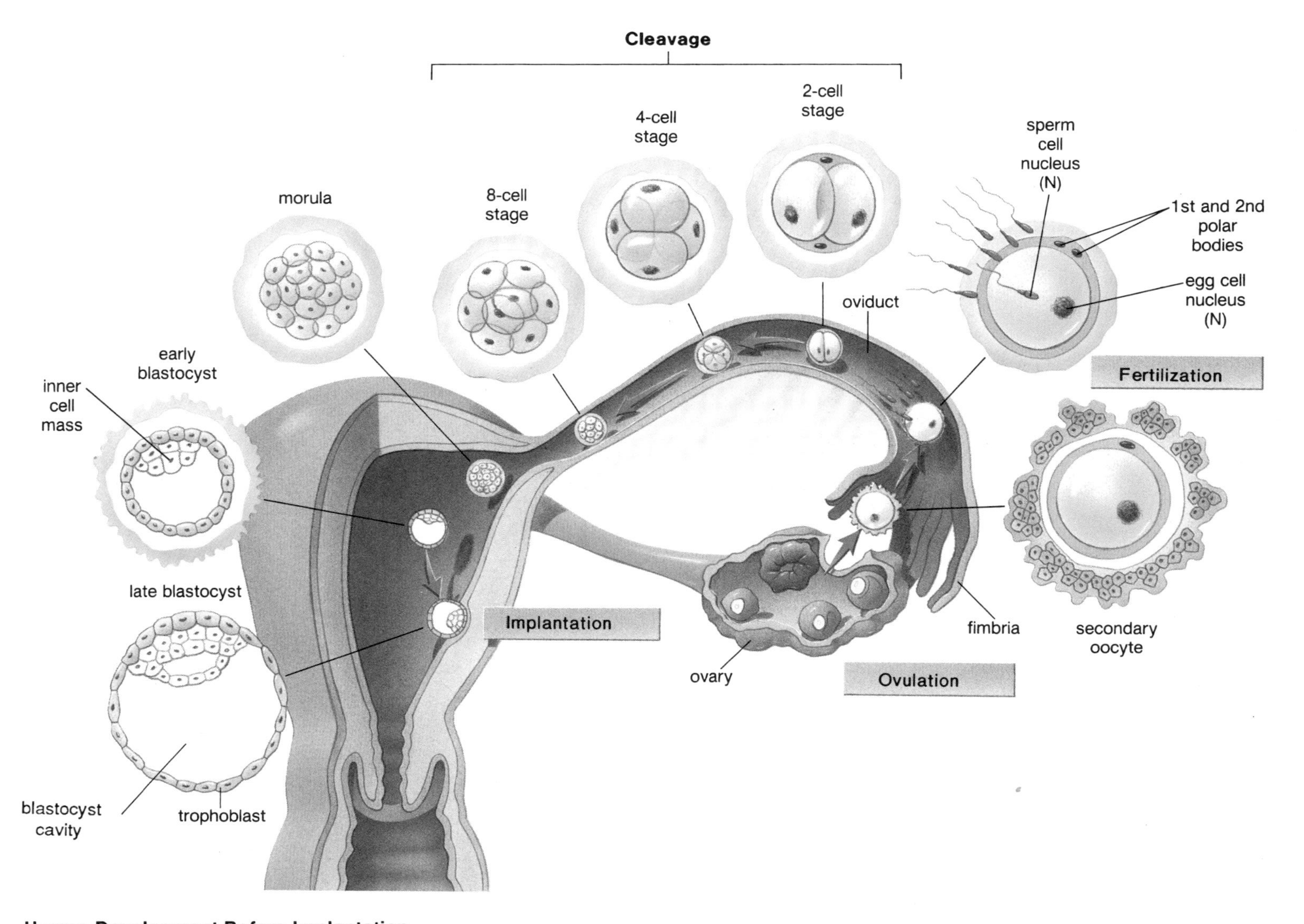
Cleavage
2-cell stage
4-cell stage
8-cell stage
morula
sperm cell nucleus (N)
1st and 2nd polar bodies
egg cell nucleus (N)
oviduct
Fertilization
early blastocyst
inner cell mass
secondary oocyte
fimbria
late blastocyst
Implantation
ovary
Ovulation
blastocyst cavity
trophoblast

Human Development Before Implantation
Figure 19.7

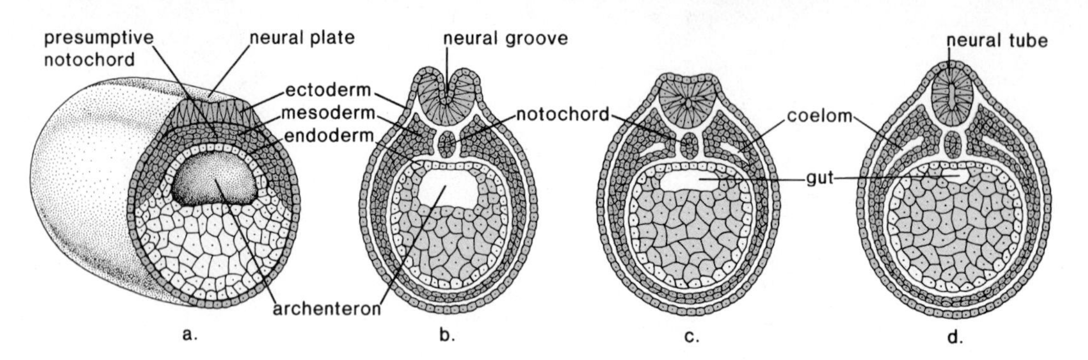
presumptive notochord
neural plate
ectoderm
mesoderm
endoderm
archenteron
neural groove
notochord
coelom
gut
neural tube
a.
b.
c.
d.

Development of Neural Tube
Figure 19.9

Cross Section of Vertebrate Embryo
Figure 19.10

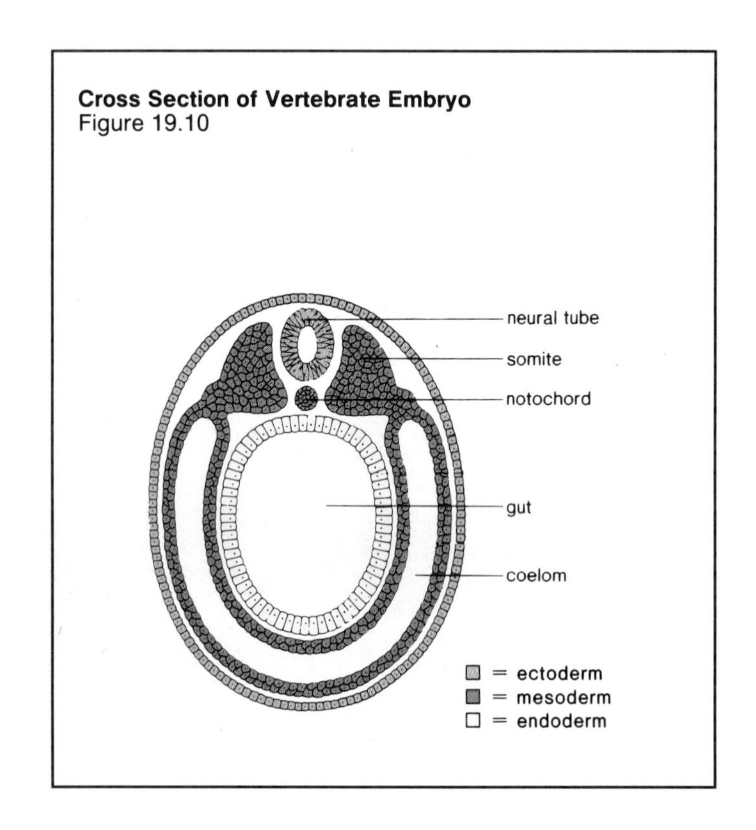
neural tube
somite
notochord
gut
coelom
= ectoderm
= mesoderm
= endoderm

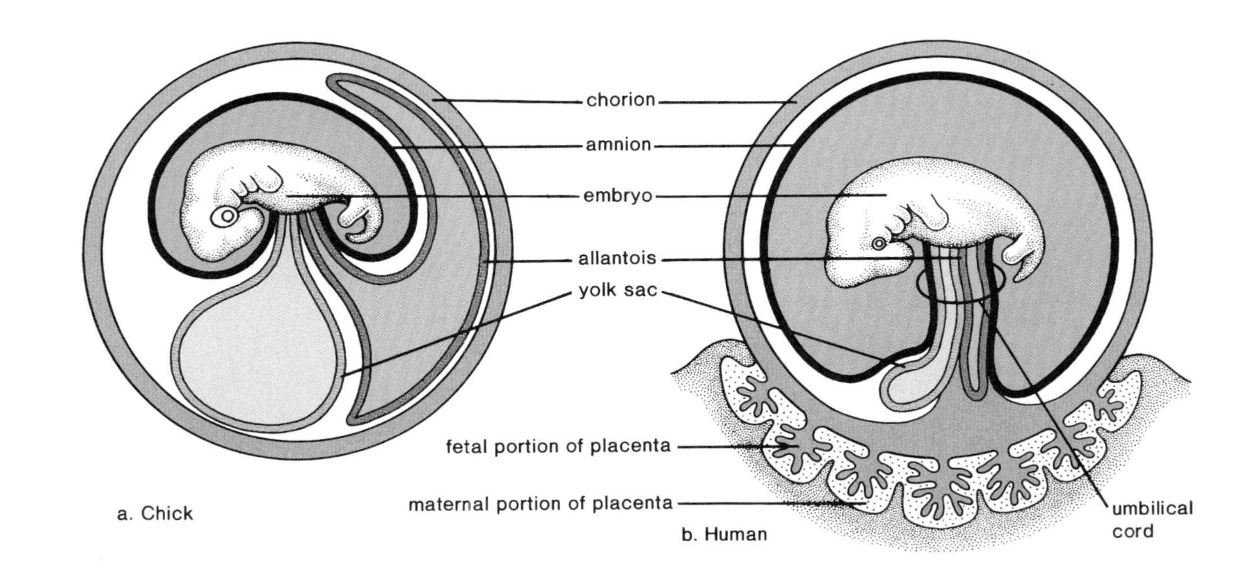
chorion
amnion
embryo
allantois
yolk sac
fetal portion of placenta
maternal portion of placenta
umbilical cord
a. Chick
b. Human

Extraembryonic Membranes (Chick vs. Human)
Figure 19.12

heat
heat
sun
producers
consumers
inorganic
nutrient
pool
decomposers
energy
nutrients
heat

Ecosystem Organization
Figure 21.3

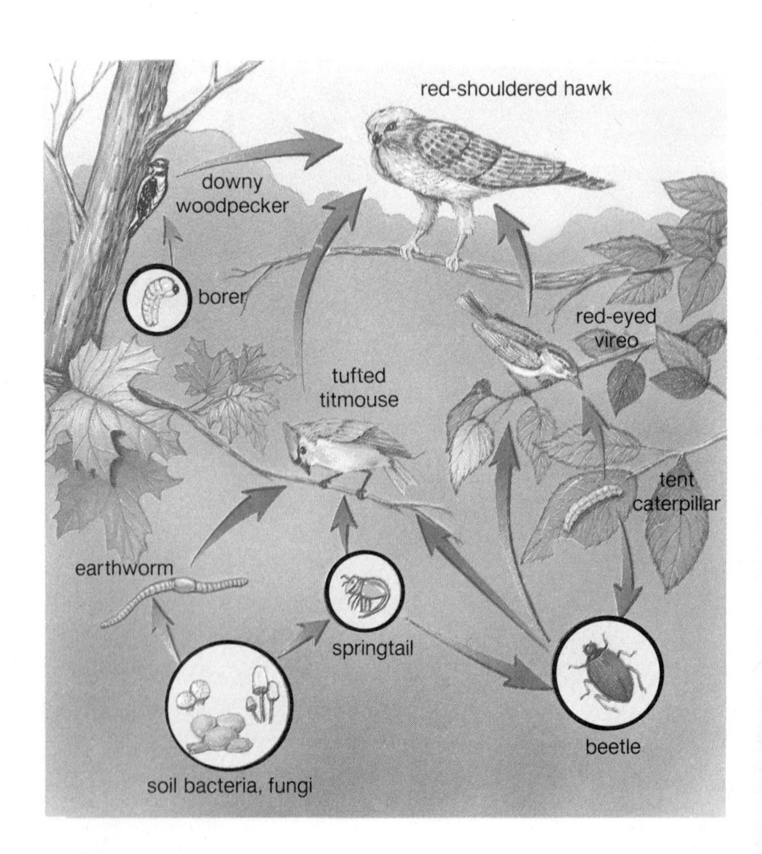
red-shouldered hawk
downy
woodpecker
borer
red-eyed
vireo
tufted
titmouse
tent
caterpillar
earthworm
springtail
beetle
soil bacteria, fungi

Forest Ecosystem
Figure 21.5

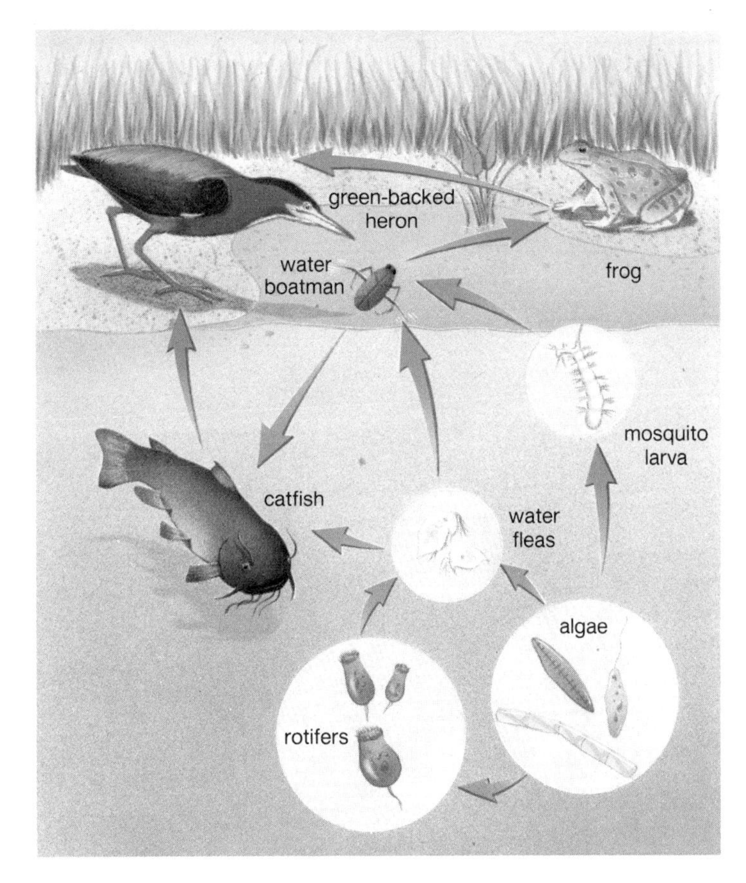
green-backed heron
water boatman
frog
mosquito larva
catfish
water fleas
algae
rotifers

Pond Ecosystem
Figure 21.6

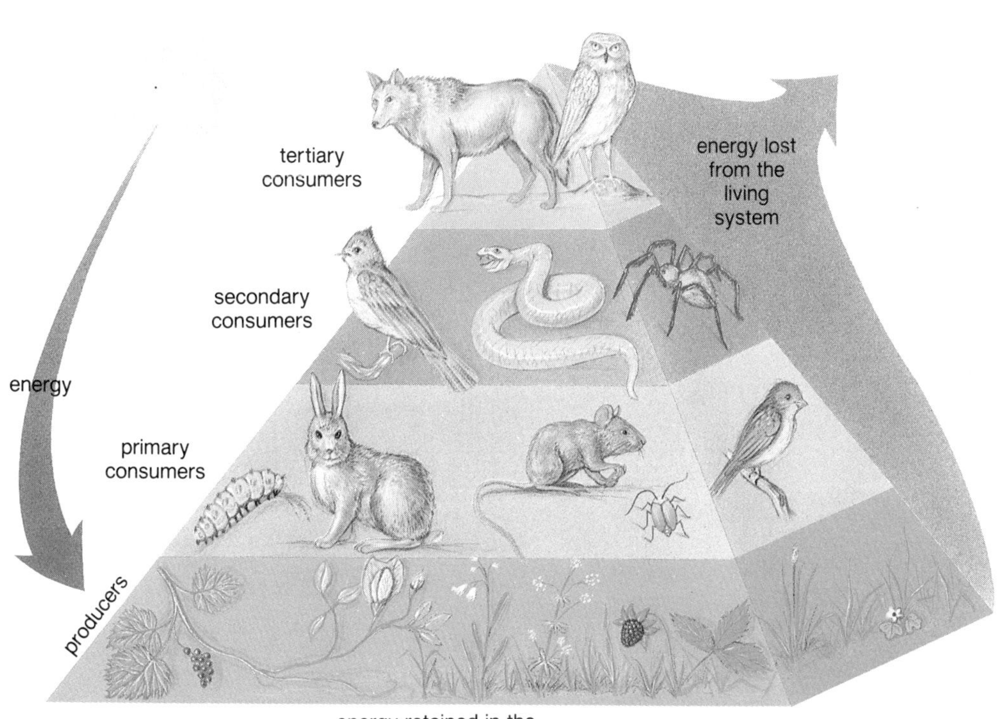
tertiary consumers
energy lost from the living system
secondary consumers
energy
primary consumers
producers
energy retained in the living system

Energy Pyramid
Figure 21.7

Carbon Cycle
Figure 21.8

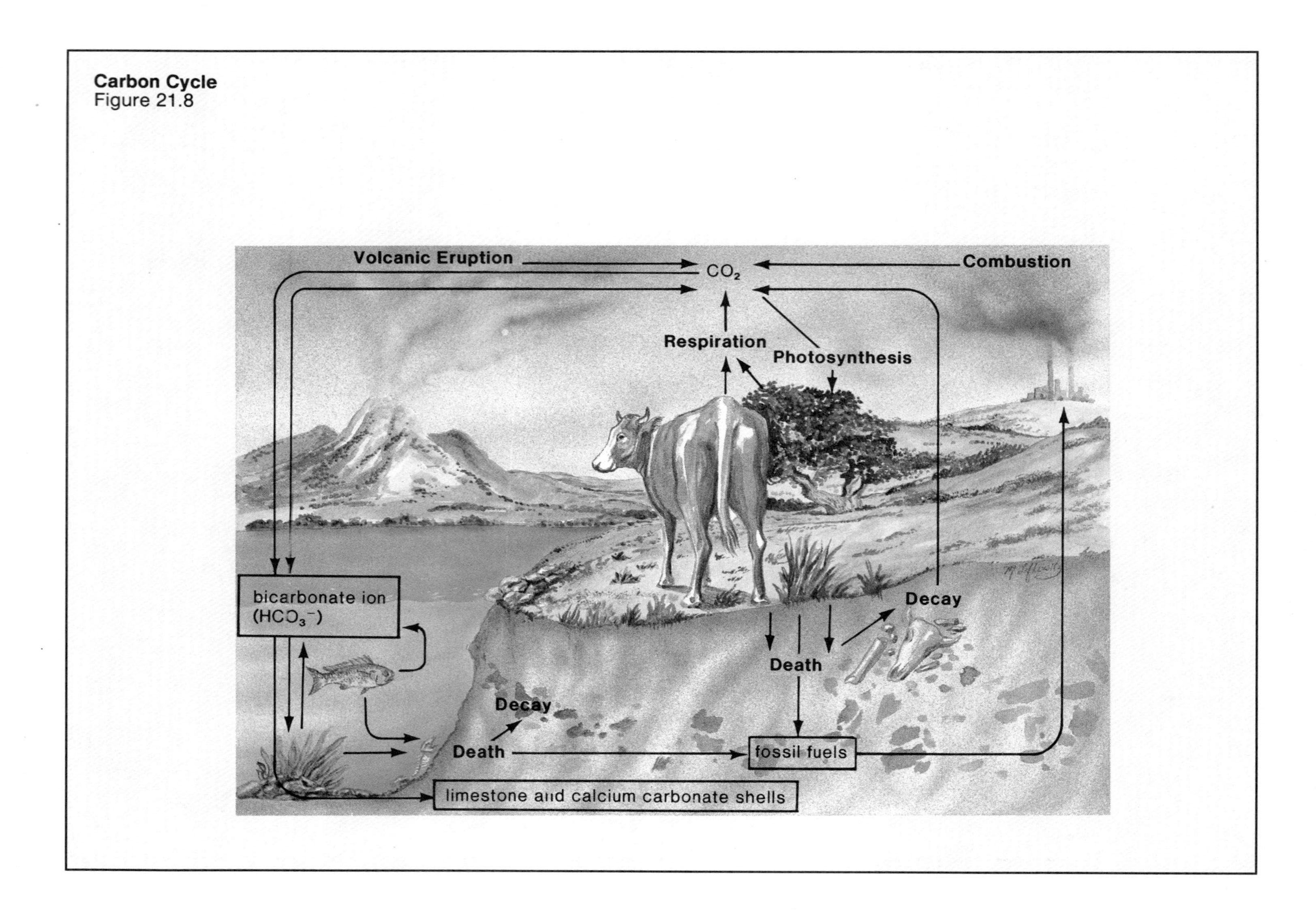

Volcanic Eruption
CO_2
Combustion
Respiration
Photosynthesis
bicarbonate ion
(HCO_3^-)
Decay
Death
Decay
Death
fossil fuels
limestone and calcium carbonate shells

Nitrogen Cycle
Figure 21.10

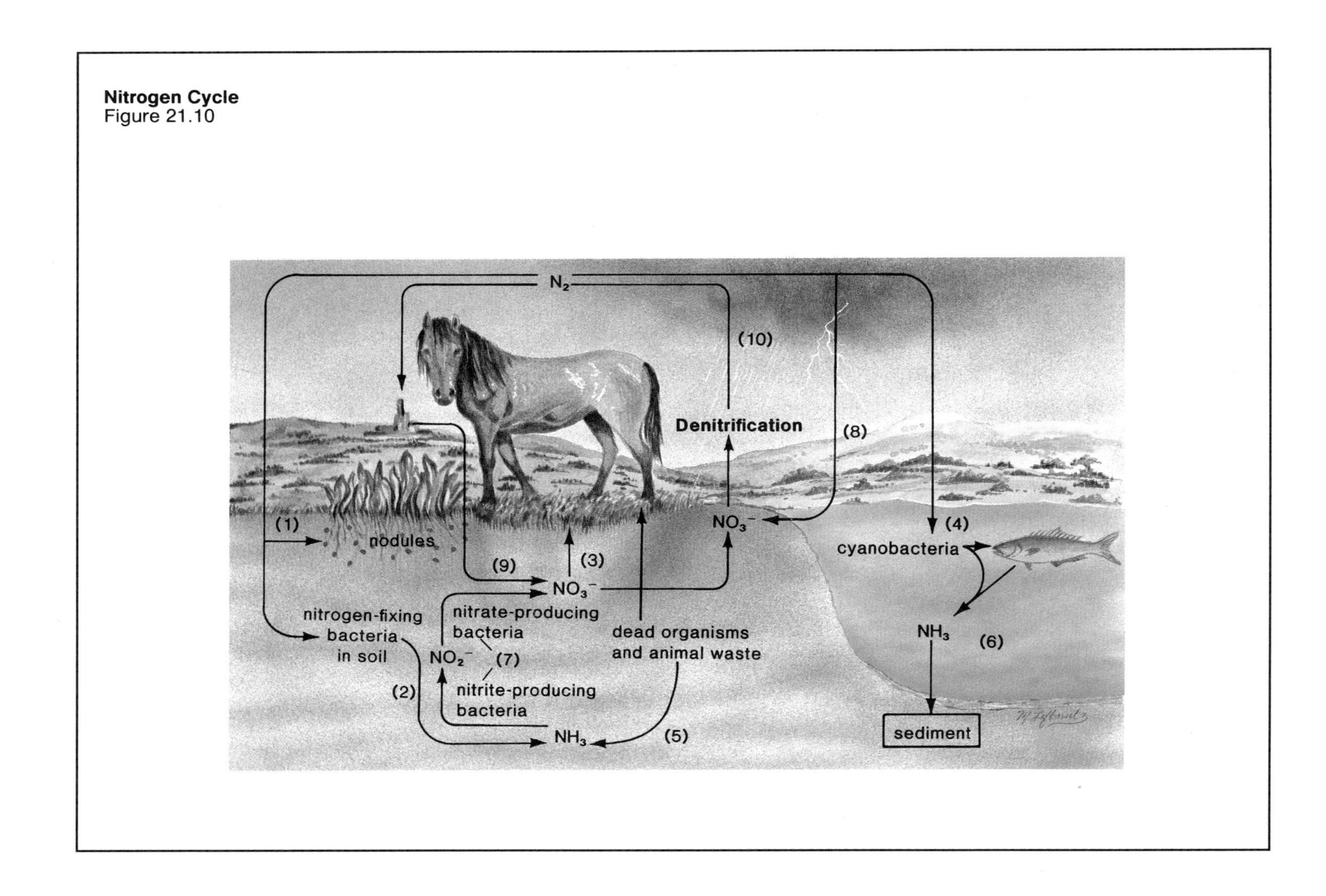

N_2
(10)
Denitrification
(8)
(1)
nodules
(9)
(3)
NO_3^-
NO_3^-
(4)
cyanobacteria
nitrogen-fixing
bacteria
in soil
nitrate-producing
bacteria
NO_2^-
(7)
nitrite-producing
bacteria
(2)
dead organisms
and animal waste
NH_3
(5)
NH_3
(6)
sediment

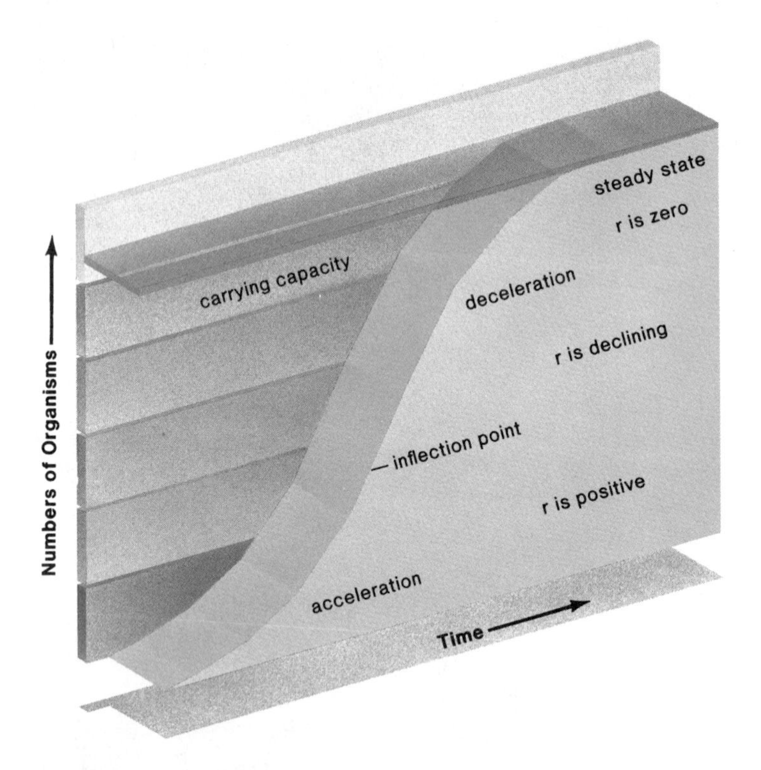
Numbers of Organisms
carrying capacity
steady state
r is zero
deceleration
r is declining
inflection point
r is positive
acceleration
Time

S-Shaped Growth Curve
Figure 22.1

CREDITS

Line Art

Fig. 12.2: Copyright © Mark Lefkowitz.

Fig. 12.12: Copyright © Mark Lefkowitz.

Fig. 15.3a: From John W. Hole, Jr., *Human Anatomy and Physiology,* 5th ed. Copyright © 1990 Wm. C. Brown Communications, Inc., Dubuque, Iowa. All Rights Reserved. Reprinted by permission.

Fig. 16.3: Copyright © Mark Lefkowitz.

Fig. 17.4: Copyright © Mark Lefkowitz.

Fig. 17.9: Copyright © Mark Lefkowitz.

Fig. 18.8: Copyright © Mark Lefkowitz.

Fig. 18.10: From John W. Hole, Jr., *Human Anatomy and Physiology,* 6th ed. Copyright © 1993 Wm. C. Brown Communications, Inc., Dubuque, Iowa. All Rights Reserved. Reprinted by permission.

Fig. 19.1: From John W. Hole, Jr., *Human Anatomy and Physiology,* 6th ed. Copyright © 1993 Wm. C. Brown Communications, Inc., Dubuque, Iowa. All Rights Reserved. Reprinted by permission.

Fig. 19.3a: From John W. Hole, Jr., *Human Anatomy and Physiology,* 6th ed. Copyright © 1993 Wm. C. Brown Communications, Inc., Dubuque, Iowa. All Rights Reserved. Reprinted by permission.

Photographs

Fig. 2.2c: © Charles M. Falco/Photo Researchers, Inc.

Fig. 3.1a-1: © Richard Rodewald/Biological Photo Service

Fig. 3.1b-2: © W. P. Wergin, University of Wisconsin, Courtesy E. A. Newcomb/Biological Photo Service

Fig. 3.5a: © Warren Rosenberg, Iona College/Biological Photo Service

Fig. 3.6a: © David M. Phillips/Visuals Unlimited

Fig. 3.7a: Courtesy Dr. Herbert Israel, Cornell University

Fig. 3.12a: © Henry Aldrich/Visuals Unlimited

Fig. 4.4c: © Myron Ledbetter/Biophoto Associates

Fig. 4.8c: Courtesy Dr. H. Fernandez-Moran

Fig. 5.5c: © B. A. Palevitz and E. H. Newcomb, University of Wisconsin/Biological Photo Service

Fig. 10.5a-1: © Brian Parker/Tom Stack & Associates

Fig. 10.5b-1: © Brian Parker/Tom Stack & Associates

Fig. 10.10b: © Biophoto Associates

Fig. 11.5a: © Wolfgang Kaehler

Fig. 12.5a: Carolina Biological Supply Company

Fig. 12.6a-1: © Arthur Siegelman/Visuals Unlimited

Fig. 12.6b: © James Solliday/Biological Photo Supply

Fig. 12.10b-2: © Michael DiSpezio

Fig. 13.4b: Carolina Biological Supply Company

Fig. 13.6a: Carolina Biological Supply Company

Fig. 13.7a: Carolina Biological Supply Company

Fig. 13.7b: © Runk/Schoenberger/Grant Heilman

Fig. 13.7c: © Runk/Schoenberger/Grant Heilman

Fig. 13.9b: Carolina Biological Supply Company

Fig. 13.10b: © Dr. Jeremy Burgess/SPL/Photo Researchers, Inc.

Fig. 14.3a-1: © Dr. Jeremy Burgess/SPL/Photo Researchers, Inc.

Fig. 14.3b-1: © Dr. Jeremy Burgess/SPL/Photo Researchers, Inc.

Fig. 15.1a-1: © Ed Reschke

Fig. 15.1a-2: © Ed Reschke

Fig. 15.1a-3: © Ed Reschke/Peter Arnold, Inc.

Fig. 15.1a-4: © Ed Reschke

Fig. 15.1a-5: © Ed Reschke

Fig. 15.3b: © Ed Reschke

Fig. 15.3c: © Ed Reschke

Fig. 17.5c: From R. G. Kessel and R. H. Kardon, *Tissues and Organs: A Text Atlas of Scanning Electron Microscopy,* © 1979 W. H. Freeman and Co.

Fig. 18.4b: Micrograph produced by Dr. John E. Heuser of Washington University School of Medicine, St. Louis, MO.

Fig. 18.15b: Courtesy H. E. Huxley

Fig. 19.4b: © P. Bagavandoss/Photo Researchers, Inc.

Fig. 21.3a: © Michael Galbridge/Visuals Unlimited